MILITARY

Field Pocket Book

TRANSLATED
FROM THE GERMAN
OF
GENERAL SCHARNHORST

The Naval & Military Press Ltd
published in association with

Published by
The Naval & Military Press Ltd
Unit 10 Ridgewood Industrial Park,
Uckfield, East Sussex,
TN22 5QE England
Tel: +44 (0) 1825 749494
Fax: +44 (0) 1825 765701
www.naval-military-press.com

in association with

In reprinting in facsimile from the original, any imperfections are inevitably reproduced and the quality may fall short of modern type and cartographic standards.

MILITARY

Field Pocket Book,

TRANSLATED

FROM THE GERMAN

OF

GENERAL SCHARNHORST,

BY

CAPTAIN HAVERFIELD,

2nd Batt. 48th Regt.

AND

LIEUTENANT HOFMANN,

5th Batt. 60th Regt.

DEDICATED BY PERMISSION,

TO

LIEUT.-GENERAL BROWNRIGG,

QUARTER-MASTER-GENERAL, &c. &c. &c.

London:

PRINTED BY VOGEL AND SCHULZE,

13, Poland Street, Oxford Street,

And Sold by

EGERTON, WHITEHALL.

1811.

TO

LIEUTENANT-GENERAL

BROWNRIGG,

QUARTER-MASTER-GENERAL TO THE FORCES, &c. &c. &c.

THIS WORK

IS

MOST RESPECTFULLY DEDICATED,

BY HIS MOST OBLIGED,

MOST GRATEFUL

AND DEVOTED SERVANTS,

THE TRANSLATORS.

PREFACE.

THE following work was undertaken by the translators, both as a useful employment for their leisure hours, and with a view to enable the younger officers of the army to profit by the result of General SCHARNHORST's experience and abilities; and it is hoped that this intention will plead in excuse for any errors, that may be found in the execution. It cannot be necessary to dwell on the merit of the original, of which the name of the author, the high estimation in which it is held by his countrymen, and the permission given to dedicate the translation of it to the Quarter-Master-General are sufficient proofs; and however the translators may have fallen short of their wishes, no trouble or expence have been spared to render it useful and beneficial to the army.

The part, treating of OUT POST DUTY, will, it is conceived, be found of particular importance at the present moment, and will convey considerable information respecting that system of petty warfare, so much practised in many parts of the Peninsula. The translation is throughout as literal as the difference in the idioms of the two languages would admit.

The part treating on PERMANENT FORTIFICATION has not been at present translated, but should the work meet with the approbation of the army, may be subsequently published.

In the original, the Plans, owing to the book having passed through several editions and other circumstances, were in many instances imperfect; but the translators have fortunately been able to correct them from the same materials, which General Scharnhorst used in framing them, and it is hoped they will now be found perfect, not only in regard to the references in the work, but also as to their accuracy with respect to the countries themselves.

As it was desirable to render the work very portable, such parts only of the APPENDIX have been translated, as were considered most useful, and the tables of coins and measures have been carefully corrected from Dubost's Elements of Commerce.

The translators now submit the Work to the ARMY, in the full belief that, if they shall thereby add to the stock of military knowledge and furnish a means of improvement to the younger officers, they shall be rendering as essential a service to the country, as their limited sphere will admit.

Although both the Officers concerned in this Work are mentioned in the Title as Translators, the Translation was in fact actually performed by Lieutenant HOFMANN, and was corrected only by Captain HAVERFIELD.

London,
July 25th, 1811.

INDEX.

PART I.

INSTRUCTIONS FOR OFFICERS OF DETACHMENTS OF CAVALRY AND INFANTRY.

PART II.

PART III.

APPENDIX.

MEASURE, RATE OF MARCHING, AND DISTANCE OF THE ENEMY.

PART THE FIRST.

Instructions for Officers of Detachments of Cavalry and Infantry.

SECTION I.

Of Patrols.

CHAPTER I.

General Rules for the Conduct of Patrols.

SUPPOSITION.

IT is supposed here, that a patrol is sent out to reconnoitre a country. The patrols, ordered to obtain intelligence of a post in rear or on the flanks of the enemy, observe the rules laid down for detachments on secret marches. See Section II.

The small patrols, sent out occasionally by outguards*, will be treated of under the head of outguards.

§. 1. *General Rules.*

a) *Horses and Men.* The best horses, and the most intelligent and active men must be selected.

b) *What Roads should be taken.* All rencountres with the enemy must be avoided, and the route of the patrol

* *Outguard.* The word *Feldwache* is thus rendered into English. The literal translation would be *Field Guard*; such a term, however, appears too general, as also does *Outpost* (*Vorposten*), nor can it mean *Picqu t* (*Piquet*), but rather a stationary guard advanced in front of particular Corps, and, therefore, may perhaps, not improperly be called *Outguard*, though not a common term in the British Service. T.

should be, as much as possible, through woods and countries, where there is least danger of being discovered: bridges should be avoided at night, and it is necessary occasionally to listen attentively, with the ear close to the ground.

c) *How to avoid being cut off.* Where this is to be apprehended, a few men should remain in the rear, or the patrol should advance in small parties by different roads.

d) *On being suddenly discovered by the Enemy.* In this case you must attack him without hesitation; unless he is considerably superior to you in numbers; if so, disperse your party, and let every one provide for himself; the man or party in the rear always retreating first.

e) *Instructions for the Men.* Except in particular cases, they should be made acquainted with their situation and the roads by which they are to return; so that, if dispersed, they may be able to provide for themselves: for this purpose, the different points of the compass, and the direction in which their own troops are posted, should be frequently shewn to them.

f) *On Returning.* A patrol, should, if possible, take a new road in returning, as the enemy might otherwise be enabled to cut it off, or lay an ambuscade upon the road by which it advanced: besides which, it gives the advantage of having reconnoitred two roads.

§. 2. *Arrangement of the March, when a Party of Twelve Men is to patrol an open Country.*

A corporal and two privates are advanced in front: the party consisting of seven privates, with the commander, follows at from 100 to 1000 paces, and two privates are at the same distance in the rear.

The men in advance should always be within sight

of the party, therefore, during the night, and in passing through an enclosed country, the distance will be less than in the day or in an open country.

Two flankers are detached by the main body to the right and left, who, in an enclosed country, or during the night, will keep as far distant as possible, so as not to run the risk of losing their party.

§. 3. *Conduct of the Flankers.*

The two men detached in advance and on the flanks, must support each other. Whilst one is searching a bush, the other must be ready with his pistol cocked, in case his comrade should be suddenly attacked. On arriving at a hill, one will remain at the bottom until the other has ascended it and acquainted him by a sign that there is no enemy in sight. If one of the flankers discovers an enemy, both immediately hide themselves, and report the circumstance to the commander of the patrol, who either places his men in an ambuscade, retires or observes the enemy from some concealed spot. If a flanker perceives an enemy advancing, he will directly fire, when all the detached men join the party.

§. 4. *March through an enclosed Country.*

In marching through an enclosed country, the flankers, both by day and night, must pass through the by-roads, which lead again into the great-road; taking care, however, if possible, never to be at a greater distance than 1500 paces from the main body.

§. 5. *March across a Plain.*

During the night, the flankers will keep such a distance as to be just able to distinguish the party, and in the day time they will not be sent out except for the purpose of examining a bush, house, or piece of low ground, &c. which may lay at some distance.

§. 6. *March in Winter, when in consequence of the Snow, &c. the By-Roads are impracticable; in mountainous Countries, where it is only possible to pass by a few hollow Ways.*

Make your men proceed by files at such a distance as not to lose sight of each other; or form small patrols of two or three men each, and let them follow each other at about 2000 paces. By this means, if the enemy intercepts the first, the second may escape without being discovered.

§. 7. *March through Woods.*

In passing through small woods, the flank parties must keep along the edge, sufficiently within the wood, not to be perceived at a certain distance. The main-body will proceed through the wood, but avoiding as much as possible the most frequented roads.

In passing through large woods, care must be taken to procure guides, and not to march by the principal roads. A support must be left behind the wood, and every enquiry must be made as soon as you have passed it, for another road, by which, in case you are cut off from the former, you may make good your retreat.

These rules must be observed both by day and night.

§. 8. *On passing a Defile.*

Before passing a defile, you must cause it to be examined by the two men in advance; during which time the flank parties will search every place on the right and left of the defile in which the enemy might be concealed.

You must also endeavour to find a road to return by without passing this defile.

Two men must be left near the defile in some concealed spot, who will give notice to the patrol, by firing, if the enemy attemps to cut it off from the defile, and that the enemy also, when pursuing the patrol, may believe that a reserve is at this place.

§. 9. *On Reconnoitring a Village or Town.*

a). *In the Day Time.* The main body of the patrol should halt at some 100 paces in front of the village; while the flankers proceed forward, and endeavour to learn from any person they may meet, whether the enemy is in possession of it; if this should be the case, they will secure the person and bring him to the commander of the party; but if he does not know any thing of the enemy, they will keep him with them, threatening him with death if they should find what he has said false.

While these two men enter the village, two others should pass round it, and through the small streets. They should enquire for the Bailiff or Constable, and search every place where it is possible the enemy may be concealed, causing all the barns, houses, &c. to be opened for this purpose; they will then give information to the main party: during this time several men should keep a good look-out in the vicinity of the village. If the commander of the patrol is a non-commissioned officer, he must procure a certificate from the Bailif, of his having visited the village.

It is necessary to be careful (particularly in reconnoitring an *enclosed* place) not to allow all the party to enter, even after having examined it; as the enemy may possibly be concealed, and then have an opportunity of seizing the whole patrol. This happened to a French patrol at Uslar, which was cut off by Lieut.-Col. Thies, under the orders of General Freytag.

b) *In the Night.* The main body of the patrol must likewise remain concealed on one side, at a certain distance from the village, and some dismounted men must approach it in a quarter where there is no road. These men will creep through the gardens, and look in at the windows, in order to discover if the enemy's troops are in the houses; or they will remain concealed until some person comes out of doors, whom they will directly seize, and presenting a pistol, demand information respecting the enemy. Emmerich, in his partisan, relates instances of this manner of reconnoitring, which is less difficult than it appears at the first view.

If no enemy is discovered, the village should be then reconnoitered in the same manner as during the day, but if the contrary, the man from whom the intelligence has been received, must be taken to the commanding officer of the party; and to procure more certain information (having learnt from the person in your possession how the enemy's guards are posted) you should approach the vedettes, and alarm them, if you cannot succeed in seizing one.

What has been said respecting the reconnoissance of a village, is equally applicable to a town.

§. 10. *Reconnoitring a Forest.*

One or two small parties must be left behind the forest, and at several places men should be sent two by two into the forest, one of whom will keep at from 50 to 300 paces distance in front of the other. If you are not acquainted with the forest, the assistance of a shepherd, woodcutter, or forester, will be necessary to you. You must march parallel to the roads, but not upon them.

The same must likewise be observed in reconnoitring a small forest, situated at a considerable distance, or part only of an extensive forest. But if a small patrol of infantry or cavalry, consisting of from three to six men, is sent into a forest, for the security of the quarters near it, the measures to be taken are perfectly different. Suppose, for instance, the situation in which the regiment of Thadden was placed at Dittersbach, in 1788. (See plate III, No. II.)

The Austrians were posted behind the ridge of hills *A*. Platzelsdorf and Neuweisbach were occupied by the Prussians. There was no outguard in the forest; the regiment had only the small guards near to it at *a*, *b*, *c*, *d*, *e* and *g*, it was necessary that the forest should be continually patroled, or the regiments might have been surprised from that quarter. How should the patrols proceed in such a case?

If a patrol does not examine the by-roads at the same time with the main one, it will be in great danger of being cut off; but in the present instance that cannot be done, on account of the extent of the forest; some other method of preventing a surprise must therefore be sought; and without doubt it may be done in the following manner: Let the patrols you send out follow each other upon the same road, at about 1000 paces distance (as described in §. 6. of Winter Patrols); if the first patrol is seized, the second will certainly escape, as the enemy cannot afterwards remain concealed, at least not at the first moment.

Let a patrol be composed of three men, and two advance abreast at some distance from, but within sight of each other, the third following so as not to lose the former; it will be very difficult for the enemy to seize the first patrol, and yet the forest will be reconnoitred. In order to discover the enemy sooner, the patrols should return by a different road from that by which they advanced. At night, the patrols should only advance about 1000 or 1500 paces into the forest, by a road pointed out beforehand.

§. 11. *Upon Reconnoitring an Enemy, Retreat.*

a) *If you fall in with an Enemy's Patrol,* which is weaker than you, conceal yourself, and endeavour to cut it off, in order to procure some intelligence by means of the prisoners.

b) *If the Enemy is superior in Number,* the patrol must not engage him, without absolute necessity; but if it cannot be avoided, he must be attacked with impetuosity, whilst a well-mounted man gallops back with the intelligence.

c) *On the Enemy advancing slowly, after having discovered you, and you apprehend that he is supported, or on his being stronger than you, his force consisting of heavy, and yours of light cavalry:* In such cases, the worst mounted men must be immediately sent to the rear, and the remainder of the party must extend and retire, firing alternately.

On arriving at a hollow way or defile, the party must be formed, to allow the horses to fetch breath, and give time for the men who are badly mounted, to gain ground to the rear.

d) The General Rule; that a patrol is never sent out for the purpose of fighting, but in order to gain intelligence, should be constantly borne in mind, and nothing ever risqued, without some trusty men being sent back. Small patrols must never on any account engage, but must endeavour to escape undiscovered.

§. 12. *Strength.*

The party will be more or less numerous, according to the number of reserves required, if two reserves are necessary, the party should consist of at least six men, twelve men or more should be employed for the reconnoissance of an extensive forest. If the object is only to examine a few villages situated in an open country, not more than three or four men will be required for that purpose.

In a long hollow way, three men are sufficient, but in this case the third man should be at a considerable distance in the rear.

CHAPTER II.

Instances of Patrols, which have been badly conducted, taken from the Seven Years' War.

THESE Examples show the necessity of precautions, and draw the attention to apparent trifles, but which are really important.

§. 13.

Hoya was very near being surprised in February 1758, by the Crown Prince, owing to the neglect of two patrols, who were ordered to proceed along the Weser to Barmen (where the Crown Prince crossed that River) and who passed the column of the Prince without perceiving it. The patrols moved along the high-road (*a*) (Pl. I, No. 4.) but the Prince advanced by the footpath (*b*), which is separated from the road by the Weser Dyke. Perhaps it was thought impossible that the Crown Prince could cross so broad a river, which had then overflowed its banks, to get into the rear of the post near Hoya : particularly as the villages along the river were occupied by cavalry. From this example may be deduced the rule, that the flanks and rear are never perfectly secure; and that detachments should always be sent into the by-roads and footpaths, and where this cannot be done, it is better to search one road after another, than to advance and leave any road in the rear, which has not been reconnoitred.

§. 14.

The Volunteers of Alsace were surprised in 1759, at Hemeln, near Munden, by Col. Freytag, in consequence of a patrol not observing the necessary precautions, and allowing itself to be cut off at Uslar.

The Colonel advanced towards Uslar secretly, through the forest of Sollingen, which lays close to it, and having received information from a peasant, that a patrol occasionally visited that place, he determined to cut it off: for this purpose he directed Lieut. Thies with a detachment, to endeavour to approach undiscovered, and place his men in ambush near Uslar, until the patrol had completely entered the place, and then if possible to seize the whole of it; this was happily executed. The patrol halted in front of the place, and sent a few men into it; the inhabitants, who probably knew nothing of Colonel Freytag's ambush, informed them that no enemy was in the neighbourhood; the whole of the patrol in consequence followed immediately, Lieut. Thies having watched until he saw the patrol had entered, fell upon them and not a man escaped.

The following rule may be drawn from this, never to enter a place with your whole detachment, however often you may have visited it before without meeting an enemy. The information of the inhabitants (particularly in an enemy's country) must not be entirely depended on, for neither threats on your part, nor the best inclination on theirs, can be of any service, when they are ignorant of the enemy's ambush.

§ 15.

Major Scheither seized a patrol near Hamm, which entered a village without having previously reconnoitred it. Major Scheither's detachment was concealed in a barn, until the patrol had passed, and had arrived in a narrow street, when he fell suddenly upon the rear, and took it together with its officer.

This may afford a reason for the following rule.

First. Always to march round the village if it is possible, and cause it to be reconnoitred by a detached party.

Secondly. If it is absolutely necessary to pass through a village, to have every barn, out-house, &c. searched beforehand, or make the detachment move through in separate parties.

§. 16.

No prescribed rule is to be neglected, even when there is not the least probability of falling in with the enemy.

That experienced officer, Captain Ewald, of the Hessian riflemen, was directed (in 1777) during the American War, to patrol at night, from Rariton Landing, to ascertain whether the defile near Boundbrook was still in possession of the enemy: his party consisted of 10 riflemen and 20 dragoons; with his riflemen he occupied a defile, which he had passed on his route, and with his dragoons he surrounded a plantation, where there had been, until then, a post of the enemy. Not finding it there, he conceived they had abandoned that part of the country, and approached, with less precaution, another plantation, which he was obliged to pass before he could reach the defile he was ordered to reconnoitre; here he suddenly received a volley from 20 or 30 men. Had he, as he himself remarks, sent for the owner of the plantation before he advanced too close to it, the whole of his men would not have been exposed to the fire, which fortunately, owing to the darkness of the night, took little effect.—See the Treatise on light Troops, by Lieut. Col. Ewald, of the Danish service, Page 133.

§. 17.

An officer of cavalry, serving with the allied army in 1760, was ordered, several days previous to the position of Sachsenhausen being taken up, to patrol with 20 Cavalry, into the vicinity of Marburg; in doing this, he was under the necessity of passing a defile, which he could by no means avoid, and which was several hundred paces in length: conceiving himself perfectly secure he entered the defile with confidence, having an advanced guard, consisting of a non-commissioned officer and 6 men, 50 paces in his front; he had not advanced more than 2 or 300 paces, when a party of hussars attacked and drove in the advanced guard upon the main body of the patrol, which endeavouring to retreat, was taken, together with its officer.

The bravest troops may be beat under such circumstances. Any thing happening unexpectedly, will disconcert the men, however accustomed they may be to danger: and if the advanced guard is driven in, that alone will be sufficient to throw the main body into confusion.

In such cases, you should at least send a few men through the defile, whilst you keep the remainder in the rear.

And if the ground continues confined, the fewer men you have with you the better.

§. 18.

Lieut. Col. Emmerick, of the allied army, was ordered, in 1759, to patrol into Franconia, for the purpose of obtaining intelligence respecting the Army of the Empire.

When he had proceeded a considerable distance on his road, having been informed by every body, and amongst others, by a catholic priest, that there was no enemy in that part of the country, he determined to feed his horses in a small forest. A detachment of Austrian hussars and the Cuirassiers of Hohenzollern, who were concealed in a village behind an adjacent wood, having obtained intelligence of it, surprised and made him prisoner.

If Col. Emmerick had fed only half his horses at a time, and had posted vedettes; or if he had halted in a place where he might have been concealed, and have seen some distance round, he would have escaped this misfortune.

§. 19.

When large patrols have to advance a considerable distance, they should move forward in small bodies, and the commanding officer should point out to them, were it only by means of a map, the places on which they are to advance.

When Colonel Polerezky, in February 1758, sent a strong patrol to the environs of Ahlen and Rethem, from Stœckendrebber, where he was posted with 300 cavalry, he was surprised in consequence of the whole of the patrol proceeding by the same road to Hœren, and thereby not falling in with the Prussians, who, at that time, had arrived at Gilten from Ahlen. If he had sent a non-commissioned officer towards Norddrebber, Gilten, Grethen, &c., while the main patrol took the road by Suderbruck, Neuhagen and Hœren, he would undoubtedly have discovered the Prussian hussars.

CHAPTER III.

Examples, supposed, of the Manner of conducting Patrols.

Suppose that a party of hussars is stationed at Mandelsloh (Pl. I. No. IV.) and infantry cantoned at Lutter, &c.; the nearest advanced post of the enemy's army being at Gilten.

§. 20. *Conduct of a Patrol sent out, from the outguard stationed in front of Mandelsloh, to Brase, Dinsdorf and the Wood near Brase.*

Pl. I. N. I.

The patrol proceeds towards Dinsdorf on the road from Mandelsloh, and on arriving near the wood, detaches two men to Brase, and two into the wood on the left: these men reconnoitre the village and the wood and fall in together again at Dinsdorf: the leader of the patrol remains with the other two men, on the road between Brase and the wood, until the return of those that have been sent out. Without this precaution, the patrol might be cut off by a party of the enemy placed in ambush. The four men who reconnoitre the villages and woods, must enquire every where if the enemy has appeared in the neighbourhood.

One of the men who remain in the rear must search all covered spots, bushes, barns, &c., having his pistol ready cocked; whilst the other watches the roads and paths to the right and left.---During the night, the patrol first advances towards the wood, and two men halt at (*a*); the other four pass through the wood by different roads, (two men always keeping together) and listen attentively for the enemy; they afterwards examine the villages in the same manner.

§. 21. *A Patrol of* 18 *Men is ordered to proceed, if possible, from Mandelsloh to Suderbruck and Norddrebber, to find out if there is an Enemy in that part of the Country.*

Pl. 1, Nos. I and IV.

After having passed Brase, Dinsdorf and the wood to the left of the village, with the precautions which have been already mentioned; six men must be sent in to search Stœcken, while the other twelve remain halted in front of the village. A non-commissioned officer must then be detached to reconnoitre the villages of Rothewald and afterwards Suderbruck; he will send four men into Rothewald, and remain himself with the other two in front of the place.

Of the remaining 12 men, one must be placed to watch the ferry at Stœcken, while the officer with 11 proceeds to Stœckendrebber; and if he does not fall in with the enemy, from thence to Nordrebber; he will send only 4 men with a non-commissioned officer into that village, and will himself halt with 7 men between the two latter places, (the ground immediately in front of Norddrebber being enclosed) ordering the bailiff to be brought to him, whom he will interrogate, and if he can procure no intelligence of the enemy, he will return by the road across the heath, (*h*) drawing in the non-commissioned officer's party, aud the man left at the ferry.

§. 22. *A Patrol of 16 Men is sent from Mandelsloh, crossing the River Leine by the Ferry at Stœcken, to reconnoitre the Country towards Esperke and Grindau.*

Pl. I. Nos. I and IV.

The reconnoissance as far as Stœcken should be made with the precautions above pointed out; a corporal and 2 men must be left to observe the country towards Stœckendrebber, and to give notice, by discharging their pieces, if the enemy cuts off the retreat of the patrol by the ferry.

After crossing the river Leine, a few men should be sent to Warmloh and Esperke; if these men and the master of the windmill situated between those places, agree in the information, that the enemy has only been seen in those environs once during the last eight days, six men must be immediately ordered to reconnoitre the woood near Esperke, and four to examine the village a second time

Not having discovered any thing, the patrol passes through Esperke leaving 2 men at the bridge as a support.

Grindau must be reconnoitred by 4 men and the wood by two, from the 12 which remain with the patrol. It is supposed that intelligence is received here, that the enemy patrols as far as Schwarmstedt and has a post at Bothmer: as the inhabitants of Bothmer go to the church at Schwarmstedt, it is probable that further information may be procured at the latter place: the patrol will therefore approach it as soon as it is dark, passing through the little wood, which lays along the banks of the river Leine.

In this wood the patrol will remain until 11 o'clock at night when a non-commissioned officer and 4 men must approach the village; the non-commissioned officer and 2 men, will dismount behind a garden, or in some concealed spot, and enter one of the houses, they will awake the inhabitants, of whom they will make enquiries, and oblige them to point out some person from whom they may obtain more particular intelligence; they will proceed in the same manner with him, and will carry him to their officer.

It is a great advantage when you are in a friendly country, to obtain information from the bailiffs, clergyman or foresters: being assured by this means that the enemy has a post at Bothmer, and that he does not extend farther, as his attention is not to be drawn to this quarter, the patrol must retire during the night by Esperke, and across the Leine at Stœcken in the morning.

§. 23. *Suppose an Army intends crossing the river Aller near Riethagen, (as the Allies did in February 1758) the Enemy is stationed (as the French at that time were) in Cantonments towards Neustadt. An Officer is detached from the former with 24 Cavalry, to reconnoitre the country between the open Marsh and the River Leine as far as Mandelsloh.*

Pl. I. No. IV.

A party of prussian hussars where detached on this service from the allied army in 1758. The officer commanding the party has no farther acquaintance with the country, than that Mandelsloh lies on the Leine about 8 miles from Riethagen, a large impassable morass, extending for about 6 miles to the right. A general knowledge of the situation of the places may be acquired from a common map.

The patrol arrives in the morning at Hudemuehle, and crosses the Aller by the ferry at that place; and information is received from the guide and the country people, that a road runs from thence, through the wood, to Buechten; and that the Aller is upon his right hand and the village of Grethen upon his left; a non-commissioned officer is then detached with 6 men to that village, with orders, should he learn that the enemy is not in the environs, to wait there the arrival of the patrol.

The remaining 18 men proceed towards Buechten, halting a few minutes on the bridge over the river Alten Aller, while the village is examined by a corporal and 4 men, and, if no enemy is discovered, they pass either round it or through it, in two parties.

From hence the village of Hœren and the borough of Ahlen may be seen, and the officer directs a corporal with two men to approach the latter place, and make enquiries respecting the enemy, from the bailif and clergyman. Towards Hœren, he detaches a non-commissioned officer and eight men, with instructions to proceed from thence to Neuhagen, halting there until he arrives with the main body, and in the meanwhile to reconnoitre that country farther to the right.

With the remainder of the patrol, consisting of eight men, the officer marches to Gilten, where he falls in with the non-commissioned officer and six men, whom he had sent there in the first instance.

At Gilten, he learns that the enemy is not in that quarter, and that a few days before he had concentrated himself more between Stœckendrebber and Mandelsloh; that the three villages of Nordrebber, Stœkendrebber and Stœcken, lay about a mile distant from each other, in an open country, along the bank of the Leine; that there is a wood in the direction of Nordrebber; and that to the left of Bothmer, is a bridge over the Leine: towards this bridge he sends a corporal and four men, who from thence join him at Norddrebber, and he procures written information from the the curate of Bothmer, whether the enemy is in that neighbourhood. He himself then moves with the main body of the patrol to Norddrebber; and from thence he has a view of the country as far to the right as Rothewald: he now determines to go to Stœckendrebber, and not to Neuhagen, he, therefore, sends one man to direct the non-commissioned officer whom he had detached to that place, to proceed to Rothewald, and, if nothing particular occurs, to remain there until he receives farther orders; but in case of alarm, to fall back upon Hœren.

The officer, having called in the corporal's party from Bothmer, now advances to Stœckendrebber, where he learns that the road to Nienburg passes by Rothewald, and that Nienburg is still occupied by the enemy; he consequently orders the non-commissioned officer at Rothewald to observe the road to Nienburg.

With the remaining 16 men, he approaches Stœcken, where

learning that the enemy is at Mandelsloh and in that neighbourhood, and as the evening is drawing in, he determines to feed his horses in a barn, which stands by itself at a short distance from the place, and then to approach Mandelsloh during the night, for the purpose of ascertaining the fact more certainly.

As soon as it is dark, he takes a guide from Stœcken, to conduct him across the heath (*b*) to the wood (*a*) without touching upon the road: and being informed by a shepherd, with whom he falls in at a sheep-fold, that the enemy has quitted Mandelsloh; in order to be more thoroughly convinced, he detaches thither by way of (*c*) a non-commissioned officer with two men, who return and confirm this intelligence.

As the officer's instructions direct him only to proceed thus far, he sends back a non-commissioned officer and one man, with his report; and at day-light proceeds to reconnoitre the country.

§. 24. *The Officer, when at Mandelsloh, as above-mentioned, receives an Order the following Day, at five o'Clock p. m. to patrol with twelve Men of his Detachment, to Nienburg, in order to discover whether the Enemy has left that Place.*

Not knowing whether the enemy is cantoned in the country through which he is to pass; he determines to execute his march during the night, and to avoid all villages. He, therefore, procures guides in the manner pointed out in §. 9. (*b*) and arrives towards morning in the environs of Zum Damm, where he keeps himself concealed, and sends a few men to the forester's house, to bring the forest-keeper to him.

By the forester he is informed, that the enemy has not quitted Nienburg, but that they have only some infantry in that place; he, therefore, quits the main road, and endeavours to approach Nienbourg concealed, in hopes of seizing a patrol, an officer's servant, or some person belonging to the garrison.

Not being able to succeed in this, towards evening he advances to the place, leaving a reserve of four men upon the dyke, that leads over the marsh to the forester's house; he makes his party extend to skirmish, and advances with one man, until he is near enough to perceive the garrison distinctly.

Immediately after which he commences his retreat by a different road across the woods and heaths, taking the forester or any man acquainted with the country, to act as a guide.

SECTION II.

Conduct of a Detachment on a March.

CHAPTER I.

Of the Articles which a Party, or an Officer, or Soldier should carry with them, when detached.*

§. 25.

Field Equipment. One tent with blankets, the necessary number of pins, and tent poles, an axe, a stewing pan, with a kettle, for every five or six men.

One bat horse is allowed for the carriage of five tents; and two additional horses to 120 men, for the blankets and camp kettles; in some armies the latter are carried by the men, and the tents and blankets on horses, or in waggons.

Provisions. The men carry bread themselves for either three or six days, (in some armies for four, and

* Whoever has observed in a campaign, how frequently these necessary things are forgotten, and the bad consequences which arise from such neglect, will find this chapter not so superfluous, as others may at first sight consider it.

others for five days.) One waggon with four horses will carry 1500 pounds of bread, therefore, if a ration consists of two pounds, a waggon of this kind will transport 750 rations, or bread for 150 men, for five days.

What the Soldier should carry about him. When a soldier is detached a considerable distance from the army, he should carry with him, besides his arms, sixty rounds of ammunition, three spare flints, a priming wire, (to clear the touch hole, a spunge, (to spunge the musquet after having fired several times), a worm, an *instrument* for taking the lock to pieces, two shirts, two pair of stockings, rags to wrap up the feet on a march, (to this and to the shoes great attention must be paid, when the detachment has a long march to perform), combs, brushes, pipe-clay, black balls, needles and thread.

What must be carried for the Horses. Forage must be carried for three days, at the rate of eight pounds of hay, and ten pounds of corn *per diem* for each horse, or corn only for six days;---and in addition to this, two spare shoes with nails for each horse, with picket poles, and nose bags.

a) *The Officer should be provided.* With explicit, and if possible written orders, as to the service he is to execute. If he is to levy contributions, or has any other duty to perform, which may subject him afterwards to misrepresentations, he must, on no account, move without instructions in writing.

He should request an explanation of any particular cases which are not mentioned in his instructions, or on which he feels any doubt :---as for instance, whether he is to maintain such a post? Whether he may retreat, if he falls in with the enemy superior in force? Whether he may risque something, if an opportunity presents itself of taking some prisoners, or executing any thing advantageously? To whom is he to report? &c.

b) Before marching the officer should make out a list of the men under his command, with the companies they belong to; without which he will not be able to send in correct returns of the sick, killed and deserted.

c) An officer cannot do without pens, ink, paper, pencils and a map. During the seven years war all reports were made upon cards, as many mistakes arose from their having been verbal. Place and date, name and rank, should never be omitted, as too frequently happens.

d) A watch is almost indispensible. By means of a map a general idea of the country may be gained at one view; and it also gives the power of ascertaining your situation, relatively to your own and the enemy's army.

A compass is of considerable advantage in shewing the direction of different places, and where your own army and that of the enemy are situated.

An acquaintance with the principal stars and constellations will be of great assistance in gaining a knowledge of the country.

No instrument is of more use to an officer of any rank in command than a telescope. It is impossible to perceive the enemy with the naked eye at a greater distance than 1500 paces, and even then it is difficult to distinguish troops, from a herd of cattle, or any thing else: but with a small telescope the uniforms may be clearly seen at 5 or 6000 paces, and infantry, cavalry, and artillery, at a still greater distance. It is of the greatest importance to see and ascertain every object as far off as possible, particularly as you may not yourself have then been discovered by the enemy.

CHAPTER II.

Conduct of Officers who command the Advanced or Rear Guards, or Flank Patrols of a Detachment.

SUPPOSITION.

Suppose it is intended to patrol a country, in order to prevent a detachment being surprised before it has time to form. The detachment, which consists of several squadrons or battalions, is ordered to discover and disperse the enemy, and conceives itself to be near him, without having received any exact intelligence of his situation.

§. 26. *General Rules.*

a) *General Distribution.* About one fourth of the advanced guard is sent on in front, and a small party is sent to the right, and another to the left, for the purpose of examining the villages, bushes, woods, &c. at a certain distance from the road.

b) *Conduct of the Advanced and Flank Patrols, and of the Advanced Guard in general.* The advanced and flank parties will send out some men in front, and to the right and left, to search the country thoroughly; they must be careful that they are not cut off from the advanced guard, which must also take every precaution, that it may not be separated from the detachment.

c) *Conduct of the Flank Patrols in general.* The principal patrols upon the flanks of the detachment will observe the same rules as the advanced guard,--they will have their advanced parties and flankers, or if they are weak, flankers only.

§. 27. *Distribution and Distance of the Advanced Guard, consisting of Cavalry.*

a) *Distribution and Distance of the Advanced and Flank Patrols.* If the advanced guard consists of one officer, three serjeants, five corporals and sixty privates; one serjeant, one corporal, and fifteen men will be advanced from 200 to 400 paces in front, and they will also send one corporal and four men forward about 100 paces, who must examine carefully all the villages they may pass through, and all the bushes &c. which lay close to the road. Each of the flank parties, should be formed of one corporal and six men, who in searching a wood, &c. will throw out two men in front, and one on each flank

In an open country, the flank parties will keep close behind that in advance. In an intersected country they must search every thing within 600 paces of the road. When they are obliged, on account of a defile, to join the advanced party, they will observe the precautions, which are recommended in §. 8, on passing a defile.

b) *Distance of the Advanced Guard.* An advanced guard should be at such a distance in front, as to enable the detachment to form before it can be attacked by the enemy, from a spot which has not been searched.

If the detachment consists of four squadrons, the distance should be about 1000 paces, varying according to the danger of its being cut off.

If the distance can safely be extended to 2000 paces, the advanced guard will be able to examine the defiles before the detachment arrives, which will not then be detained.

c) *Distribution and Distance of the Flank Patrols.* If each of the principal flank patrols (that is the flank patrols of the detachment) consists of fifteen men, four men should form the advance and three the flank parties. The distance of this patrol from the detachment, when the latter consists of four squadrons, should be from 800 to 2000 paces; this distance should be more or less according to the strength of the detachment. The patrols must conform to the roads, which often require them to keep the same road with the detachment for some time. There should never be any impassable ground between the detachment and the flank patrol, if the latter meet with any thing, they must immediately close in: otherwise they will march abreast of the detachment.

§. 28. *Distribution of the Advanced Guard, composed of both Infantry and Cavalry.*

Suppose the advanced guard to consist of forty infantry and twenty hussars----ten hussars should then be sent in front, and five on each flank, during the day---- but in marching through a wood, or in a dark night, these parties should be formed of infantry.

It is to be observed, that the hussars in front and on the flanks should follow the same rules as are laid down for an advanced guard of cavalry.----The main advanced guard should never be at a greater distance than from 300 to 1000 paces from the detachment.

§. 29. *Distribution and Distance of the Advanced Guard and Flank Patrols, consisting of Infantry only.*

The distribution of the advanced guard and flank patrols, will be the same as the cavalry; but in passing through woods, their distance from the detachment should not exceed from 400 to 500 paces, in an open country they will keep close to it.

§. 30. *Rules to be observed on the March and on meeting an Enemy.*

1. *Duty of the Flankers.* a) The cavalry flankers must have their pistols cocked, those of the infantry will take off their hammer caps.

b) All flankers should enquire of the country people, herdsmen, &c. whether any of the enemy's troops are in the neighbourhood.

c) They should look frequently to the party, so as not to lose it.

d) When one of the flankers is examining a small village, bush, &c. his comrade will remain in front of it until he has finished.

e) They should ascend all small hills and rising grounds, from whence they can have a view of the country.

2. *Rules to be observed on discovering the Enemy.* a) On falling in with the enemy unexpectedly, the flankers will discharge their pieces, and report the circumstance to their non-commissioned officer, who will also report it to the officer, by whom it will be forwarded

to the detachment, and who will endeavour to ascertain the fact more perfectly, being careful not to shew himself if he has not been already discovered.

b) If the enemy suddenly appears from an ambush, the advanced guard must instantly attack him on all sides with impetuosity; by which, time may perhaps be gained for the detachment to form.

c) An advanced guard or flank patrol, being pursued, must never fall back direct upon the detachment. This would be a serious fault.

§. 31. *Rules to be observed in passing a Village or Defile.*

a) The advanced guard remains in front of the village, until intelligence is received that it has been reconnoitred by the advanced and flank parties, who must be careful to search every barn, &c. or they may be exposed to the accidents pointed out in § 14, and 15.

b) The advanced guard must not enter a defile, until the party in front has arrived on the other side, and those on the flanks have reconnoitred the ground to the right and left. If any small roads lead into the defile, two men must be sent into them, one of whom will return with the intelligence, while the other remains at some spot from whence he can have a good view of the road.

§. 32. *In passing a Wood.*

The advanced party sends some single men several hundred paces in front, as already mentioned, who will examine every thing to the right and left of the wood.

In thick woods the parties on the flanks of the advanced guard, will be from 200 to 500 paces distant, and their flankers extending still farther; if there is much underwood, these parties will move upon such roads as there may be to the right or left. The advanced party sends its flankers in different directions as the ground will admit.

§. 33. *Guides.*

Officers commanding advanced and flank parties must make themselves as well acquainted as possible with the roads they are to move upon; and must obtain a list of the places they are to pass through. They must also procure foresters, herdsmen, &c. as guides; and must endeavour continually to ascertain their situation by means of the compass and map.

§. 34. *Example of the Disposition in particular Ground.*

Two squadrons are detached from an army stationed at Ahlen (Pl. I. No. IV.) towards Mariensee, in order to obtain intelligence, and to disperse the enemy's patrols. The march of this detachment took place at the time when the allied army, in February 1758, crossed the Aller near Ahlen, and drove the French out of Hanover along the river Leine. The advanced guard consists of half a squadron, and is formed as pointed out in § 27. The march is directed along the Leine, first by way of Grethen, and from thence to Gilten and Nordrebber. The advanced guard does not enter the villages until its own advance has passed through them, and sent back a corporal to acquaint them of it. The party on the right flank of the advanced guard, does not pass through Gilten, but crossing the wood between that place and the morass, moves along the roads laid down, leaving Nordrebber to the left. The other flank party proceeds along the road between the river Leine and the village of Gilten, until it arrives in front of the bridge near Bothmer, across which it sends a few men into Bothmer to enquire after the enemy, from thence it continues its route to Nordrebber, where it joins the advanced guard, which now advances towards Stœckendrebber.

The flank party on the right moves on abreast of the advanced guard, along the road which runs between the windmills.

From Stœcken the advanced guard proceeds to Amendorf. The advanced party with its flankers reconnoitring the wood.

The flank party on the side of the river Leine, moves towards Dinsdorf, Brase and Mandelsloh. The party on the other flank taking the route between the heath and the fields. The principal flank parties, consisting of thirty cavalry, which have hitherto remained with the detachment, now form a separate patrol on the righ t flank towards Lutter, Bevensen, Dudensen, and Hagen.

One of the flank parties of the advanced guard moves along the bank of the river Leine; the other, by the mill towards Buehren until it joins the advanced guard near Mariensee.

When the Prussian hussars, in February 1758, reconnoitred this country, they advanced no farther than Gilten, where they were informed that a regiment of French hussars was stationed at Stœckendrebber, which they surprised *.

§. 35. *Conduct of an Officer of Cavalry commanding a Rear Guard.*

1. *Distribution.* If the rear guard is formed of one officer, one serjeant, one corporal, and twenty privates, the main party consisting of one corporal and ten privates, follows the detachment at 500 paces distant; 2 flankers are sent out to the right and left, who, in an enclosed country, will extend as far as they can without losing the party; the serjeant follows with six men at from 200 to 400 paces distance, having two men about 100 paces in his rear.

* It is stated here in the German, "that the country is not represented quite correctly in the plan," but this is omitted, the plans having, as mentioned in the preface, been corrected from the original materials. T.

2. *Rules to be observed when pursued by the enemy.* When a rear guard is pursued, it should be formed in two divisions, which should keep near each other, and one third should extend and act *en tirailleur.*

Of these *tirailleurs* every two men remain and act together, firing and retreating alternately.

The same must also be done by the parties, when one fronts, the other retreats; they are formed in single rank, and when hard pressed, the party which is fronted (if their men and horses are trained to it), will fire with their carabines.

See the remarks respecting skirmishers, in the section on conduct in action.

§. 36. *Conduct of Advanced Guards, Flank Patrols, and Rear Guards, under some particular Circumstances.*

1) In foggy weather, and during the night, the distribution is the same as in the day; but the distance of the parties from each other, and of the advanced guard from the detachment is less, and the communication between them must be kept up by single men, that the rear may not lose those in front.

2) When the object is to *surprise* an enemy, the advanced guard must be close in front of the detachment, and the flankers in reconnoitring must keep themselves concealed. See the section on surprises. The flank parties and flankers should remain near to the advanced guard, or should not quit it all.

3) When a detachment does not exceed one squadron, or has a fixed destination, for instance, to occupy a distant place, as also in marching through a country, where there is no chance of meeting an enemy, it will be

unnecessary to adopt all the above precautions. The flank parties, which are very harassing to the men, may be dispensed with. However, it is always better to be too prudent than not sufficiently so.

The rules above laid down, are applicable to cases, where the greatest caution is requisite.

CHAPTER III.

Conduct of a Detachment on a secret March.

SUPPOSITION.

If the object is to attack, surprise, or reconnoitre a post, to pass through a country occupied by the enemy, in order to arrive at a fortress in possession of your own troops, or to join a corps of the army, or to destroy the magazines in the enemy's rear:---all marches undertaken with these views, are denominated secret marches.

§. 37. *General Precautions against being discovered.*

a) *Of the Time of the March.* The detachment should march generally by night, remaining concealed in the day, or marching only through woods, and avoiding as much as possible all villages and inhabited places.

b) *Provisions.* If the forage, which the detachment carries with it, is not sufficient, what is requisite must be procured from lone farm houses; the detachment remaining concealed whilst it is received. The people who deliver the provisions or forage should be told, that, if they betray you, their treachery will be severely punished.

It is better if the persons sent to procure the forage are disguised, or at least dressed in great coats. Secrecy is of the utmost importance in the execution of this service.

c) *Guides.* You must endeavour to procure people from their houses at night to act as guides (as in §. 9. *b.*) they must not be discharged until there is no possibility of their betraying you; and even then they must be threatened with death and the destruction of their houses if they are found treacherous: at the same time, they must be kindly used and well remunerated. If the affection of one of these people is gained, he may be almost as useful as a spy in that country; information is often drawn from them by kindness which they would otherwise have concealed.

d) *Of the Country People who meet the Detachment accidentally.* Where the detachment is situated in the midst of the enemy's quarters and posts, every countryman who may accidentally discover it, must be detained, and not released without threats of the most severe revenge upon himself and family if he betrays it.

§. 38. *Disposition of the Troops on the March.*

a) *Strength of the Advanced and Rear Guards.* The strength of the advanced guard must be such as to admit of its sending out three parties of several men each, one in front and one on each flank. When the detachment is weak, one-third must be employed for this services; but in a strong detachment, one-fourth or one-fifth is sufficient.

The rear guard, having only to keep up the stragglers need not be composed of more than six men.

b) *Disposition when the Detachment consists of either Infantry or Cavalry only.* The advanced guard and flank patrols must keep near the detachment to avoid being discovered; but they cannot be entirely dispensed with, as you might then be in danger from an ambuscade, or might be attacked by a party of the enemy before you had time to form. In marching through woods, and during the night, the advanced guard should fol.ow its advanced party at about 100 paces, and the detachment should also be about 100 paces in rear of the advanced guard.

The advanced guard has two flank parties, which remain with it in an open country during the day; but when the ground is intersected by woods, hedges &c, both by day and night they must examine every thing within 200 paces distance.

During the night, and in passing through woods, the flank parties move abreast of the advanced guard, at from 50 to 100 paces distance, as the ground will admit. In passing a defile or village, the rules must be observed which have been already prescribed in treating of patrols; but with this difference, that you should endeavour to pass the village secretly during the night, which may be effected if you are not obliged to move through the principal streets.

c) *Infantry and Cavalry.* In thick woods, steep mountainous ground and during very dark nights, the advanced guard &c, should be furnished by the infantry; in other cases, this service should be performed by the cavalry, who are more adapted to reconnoitre a country, to seize any persons who may discover the detachment, and to surprise the enemy, or keep him in play until the infantry can throw themselves into enclosed ground.

§. 39. *Rules to be observed while feeding the Horses on the March, halting or being obliged to seek Shelter from the Weather during the Night.*

1) The advanced and flank parties must be particularly alert, and during the night some of the men must listen attentively, laying their ears close to the ground; the instant they discover a party of the enemy, they must conceal themselves to enable the detachment either to avoid it or to seize it, if they have an opportunity.

On these precautions depends the success of the enterprize, and it was owing to them, that, that on Hoya and many others of the same nature have succeeded. During dark nights, it will be found advantageous to send an officer and a few men dismounted about twenty paces in front of the advanced party, to listen frequently and attentively. If the detachment consists of infantry, the men should lay down on the ground and allow the enemy's patrol to pass; if of cavalry, it must halt, and the advanced party must endeavour to get into the rear of the patrol; if of infantry and cavalry, the advanced guard being cavalry, and the party in front of it infantry, the latter must lay down concealed, and thus cut off the retreat of the patrol, calling out to them that they are prisoners, whilst the advanced guard rushes upon them.

When the nights are clear, the detachment should,

if possible, conceal itself on one side of the road, as soon as they perceive any thing at a distance; this may be generally done, particularly if the party consists of infantry.

2) It is to be understood, that loud talking and smoaking during the night must never be permitted, when the detachment arrives near the enemy, whom it is intended to surprise: and it must always be supposed that the enemy is upon your route.

The horses should never be fed in a village, nor in a large wood; but when doing so the detachment must be clear of the road, which should be carefully watched: it is better, if possible, to feed in a small wood, where you are concealed, and at the same time are able to see for some distance round; but at any rate, vedettes and small parties should be posted to give the immediate alarm, in case of danger; and only half the horses should be fed at a time. (See §. 18.)

When you wish to rest during the night, to obtain shelter from the weather, you must enter some lone house, which has been previously examined by a few men; but in this case every body in the house must be carefully watched, lest you should be betrayed. You must not enter the house until it is dark, and endeavour to have the court yard locked, and its entrance guarded by a few men, who will be constantly on the alert. Part of the detachment should remain mounted, or should not quit their horses.

§. 40. *Intelligence of the Enemy.*

This is to be obtained from country people, who must be taken secretly at night from houses, which stand apart from the village. (See §. 9. b). You must make these people point out the clergymen, foresters, herdsmen, guides, &c. who may be able to communicate any intelligence of the enemy, and who must be laid hold of in the above manner; they should not be taken to the detachment, if it can be avoided, but should be detained and threatened; they may also be useful as guides.

§. 41. *General Rules to be observed, in order to avoid being surprised.*

a) The history of the seven years' war furnishes many instances of detachments, engaged in secret marches, having been surprised by an enemy concealed with some other intent. On a secret march, therefore, it should always be supposed that the enemy is concealed somewhere in the country which the detachment is to pass, or that he may himself be employed in a secret march: it is therefore necessary to be constantly prepared for action.

b) If, on a secret march, you do not change your stations frequently and suddenly, and conceal the direction of your march; if you remain too long in the same place, and return on the same road by which you advanced; there is every probability you will be surprised and taken; as the enemy, although you do not know it, will always have intelligence of your movements, which you must render useless to him by preventing his discovering you.

You must observe your men constantly on night marches, particularly when your object is to surprise a post: if a man deserts, you are betrayed, and then you have no alternative but to retire, or change your direction immediately.

In long marches, it will be well to feign an intention perfectly different from the real object in view.

CHAPTER IV.

Disposition of a Detachment under particular Circumstances.

§. 42. *Explication.*

The disposition must depend on the object of the march, the description of troops that compose the detachment, and the nature of the country through which it has to pass.

a) *When the Object is to discover the Enemy.* The same rules must be observed with respect to the advanced guard and flank patrols, as are laid down in Chapter II. The advanced guard should consist of one-third or one-fourth of the detachment, or at least should be sufficiently strong to furnish an advanced party and several flankers, and that the guard should still consist of a few men.

b) *If a Detachment is to attack and dislodge a Post inferior in Numbers, and can get into its Rear, and surprise it.* The measures to be taken are the same as above, except that the advanced and flank parties do not extend so far, and the flank patrols are dispensed with.

c) *If the March of the Detachment is in Rear of the Army, where it is quite secure from Attack.* The detachment will have an advanced guard, which will send out its advance and flankers, but no parties on the flanks; it is only in woods that some flankers need be sent out. But in this instance a great deal depends on circumstances, and on the discretion of the commanding officer.

d) When the route of the detachment is in the rear, where no enemy is to be expected, it is only necessary to send flankers to spots near the road, where it is possible he might be concealed.

e) *If the Detachment consists of Cavalry and Infantry.* The advanced and rear guards will be furnished by the infantry in woods and during dark nights; but on plains, in a country that is intersected by hedges, or that is not very steep and mountainous, and when the nights are clear, the cavalry are better adapted for this service.

f) *In the March of a small Detachment the Men should not be constrained too much.* The men may always be allowed to march by files, when the roads are bad; but the divisions must not be permitted to mix together, which produces great confusion, even in small bodies, if an alarm takes place, particularly when the detachment is passing a defile, or in a dark night, rainy weather, &c. but this will unavoidably be the case in the night if irregularity is allowed in fine weather, and in good roads.

g) The rules for attacking or surprising a post or detachment, have already been given in Chapter III.

The regulations for the conduct of a detachment ordered to reconnoitre the enemy, or to patrol a country, &c. will be noticed in the paragraph treating on that subject.

SECTION III.

Of surprising and seizing Posts.

CHAPTER I.

General Rules.

§. 43. *The principal Points to be considered relative to a Surprise.*

a) Knowledge of the enemy's posts.

b) Situation of his guards.

c) Roads of the patrols.

d) Distance of the other posts.

e) Knowledge of the roads by which it is possible to get into the enemy's rear*.

It is true that you can seldom acquire all the information that is here said to be necessary, but you must take into the account, that in these cases, the attacking party has a great advantage, and that confusion and fright, unavoidably attend men who are surprised.**

* If it is possible to turn the enemy, you will generally succeed in surprising him ; and it almost always is possible, if you are acquainted with the country, and have good guides. The most vigilant and experienced officers sometimes neglect their rear, and there are always foot-paths, or some other means of passing undiscovered. Even the famous partisan Fischer was surprised near Wettern, in 1759, owing to a neglect of this kind. The seven years' war furnishes many instances of surprises succeeding in consequence of making a detour only.

** Lieutenant-Colonel Ewald, in his treatise upon light troops, says, " I have only once been witness to a false alarm, where a party conceived itself

However, if you can obtain no certain intelligence of the enemy, and of the post you purpose to attack; if he has been put upon his guard, or has been roused to vigilance, by former attempts of this nature, and if you cannot get into his rear, a surprise will seldom succeed.

§. 44. *Circumstances which are favourable to a Surprise.*

a) Nothing will favour the success of a surprise more than the difficulty of its execution, as this renders the enemy careless, and particularly the soldiers, who always lose their confidence when an enemy appears unexpectedly, and confusion then inevitably ensues. Several instances of surprises of this nature will be found at the end of the present section. The enterprises of Major Scheither, near Homburg, in 1758, and Ordingen, in 1759, succeeded merely from the French conceiving that the enemy was at a considerable distance, and that the Rhine secured them from attack.

b) The weather may likewise cause the success of a surprise, which under other circumstances would be impracticable. When the weather is stormy and in heavy rain, it is impossible for a party acting on the defensive to guard every point so completely as when it is fine.

---

surprised: it is hardly to be imagined to what a degree fright operates upon men awaking from sleep. It was during the campaign in Pensylvania, the Hessian and Anspach riflemen were stationed in a wood, and lay upon their arms, ready to march at a moment's notice. Several shots were fired at one of the picquets, which occasioned a loud outcry among the inhabitants of a neighbouring plantation: suddenly a voice called out: Run, we are surprised! The whole corps immediately dispersed; it was above an hour before they could be brought together again, when it was hardly possible to convince them that it was a false alarm." A party which is surprised will take hedges and bushes for troops; any circumstance however trivial, or any appearance, agitates them, and increases the confusion.

Fatigue will frequently make even the most cautious relax their vigilance. Many surprises were happily executed during the seven years' war, owing to fog, rain, or snow, which would not have succeeded under other circumstances. It is generally as easy to turn an enemy's post in bad weather, during the day, as at other times during the night*.

c) If you can put the enemy off his guard, by drawing in a post, or retiring gently, and then advancing again rapidly and secretly by a different road, you will have every probability of success.

d) By continually harrassing, you will render the enemy bold and careless, at least to such a degree, that in the first moment he will not regard the attack.

e) If you receive information that the enemy intends to occupy a post in an enclosed country, you may generally succeed in surprising him, by placing an ambush in rear of the post. A French partisan, with a detachment of 150 infantry, from the army of Soubise, in this manner surprised four squadrons of hussars, and a batalion of light infantry, in August, 1761. He remained concealed until night, and although the hussars had their horses bridled, and the infantry lay with their arms close to them, a considerable number were killed and taken.

A similar surprise was executed during the American war, which Colonel Ewald mentions, in his book already noticed, in the chapter treating of surprises.

f) If it is possible to approach undiscovered near an

---

* Lieutenant-colonel Ewald, in his Treatise on light troops, says, "I recollect instances on the most dangerous posts, and in bad weather, where the least neglect in the centries might have cost them their lives, that I have approached and remained almost close to them, for a short time, without being perceived!

enemy, who has just arrived at some spot, and has not yet posted his picquets; or having been there for some time, has posted them badly; or who, while feeding his horses on a march, has taken no precautions; success may reasonably be expected, even in the day. The town of Nordheim, which is walled, was surprised in this manner, in the middle of the day, (1760), by a party of Hanoverian riflemen, owing to the French posts not commanding a view of the country round them. The seven years' war supplies many instances founded upon those faults.

Secrecy is the principal Point to be observed. In all surprises the chief object is to keep the design secret: if the enemy, from your preparations, by means of his spies, or through country people, or from his post having been reconnoitred, is led to suspect your intention, he will be upon the alert, and, perhaps, prepare an ambuscade.

Experience shews that designs of this nature are often discovered in a manner that is perfectly incomprehensible, even when the greatest caution is observed: and it may truly be said, that in such cases you are never sure of success*.

* When Major Scheither was in winter quarters at Dulmen, in 1759-60, he could never undertake any enterprize, of which the enemy did not receive previous information; he, therefore, for some time made a practice of drilling his men on the outside of the town, and one day moved off suddenly from his exercising ground; another officer occupied all the roads, by which information could be conveyed to the enemy, with small parties; and to make his men more attentive in seizing every body, he gave out that he had received information that a large sum of money was to be sent to the enemy, and was to be carried by several single persons.

§. 45. *How Information is to be obtained of the Enemy, and of the Situation of his Posts.*

This information is to be obtained. First, by means of persons who will set out before you, in disguise, and wait your arrival at a certain spot. Second, or it may be gained from persons coming from the neighbourhood where the enemy is stationed, or who have resided, or were born there : such people are generally to be found in a country some miles distant from the post.

Good information is sometimes received from deserters, but it must not be too much relied upon, as it is often false; and deserters are not unfrequently spies. Your information must either be procured prior to your setting out, or before your march, in which case you can only move slowly.

§. 46. *Of the March and Time of Arrival.*

You must approach the post you purpose to surprise, by secret marches*; and endeavour so to regulate your movements as to arrive at the point before day-break, or if it is a small detachment that you intend to attack, soon after midnight.

Immediately after executing your design, you must fall back a considerable distance; and if it is probable that the post will be speedily and powerfully supported, you should endeavour to arrive there some hours before day-break, so as to be able to withdraw again during the night.

---

* Lieut.-colonel Ewald does not approve of having any flank patrols or flankers, even if they are at as short a distance as possible. He thinks the advanced guard should be close to the detachment, with an officer on foot, about fifty paces in front, who is to listen attentively, and act according to the rules laid down for secret marches.

If the enemy relieves his guards at day-break, you should make the attack three hours before that time.

§. 47. *Precautions with Respect to the Defiles which it is necessary to pass.*

The defiles, through which the party must retire, should be occupied in order to secure your retreat. If the enemy pursues you, he will hesitate at the sight of fresh troops, by which you will gain time to pass the defile: and, if it is his intention to cut you off, this is the point at which he will endeavour to effect it; your party will therefore be able either to prevent it, or give you timely notice of it, when you may take another road.

§. 48. *Precautions relative to Succours advancing to the Enemy.*

Some of the party should be detached to oppose the troops advancing to the enemy's assistance. The defiles through which they can pass should be occupied or rendered impassable. Unless this is done, you may yourself be surprised.

These precautions are however unnecessary. First, if the surprise is decided on instantaneously, and Secondly, if the enemy cannot arrive in time to support his post.

§. 49. *Disposition to preserve Order in the Attack.*

a) *Of the Division and Disposition.* The attack should be made in several divisions, which are to act separately, but to conform to each other By this disposition, the confusion which easily takes place at night, will be avoided.

The infantry should be formed two deep, that the first rank may not be annoyed by the fire of a third.

b) *Of the Description of Troops.* When it is necessary to remove any impediments, either natural or artificial, and when the nights are very dark, the infantry should be formed in front: but, when a village, an unfortified town, or a detachment stationed in the open country, is to be attacked, where it is of the utmost importance to take advantage of the first moments; the cavalry should lead.

c) *Sign.* A sign should be given, by which the men may recognize each other; it consists generally of a word, and a piece of white paper worn on the hat or arm; every individual should be in possession of it.

d) *Rendezvous after the Attack.* One or two places should be fixed upon, to which the troops, on a given signal, must retire after the attack. This signal should be the beating or sounding of a familiar march, and which should be continued, even after the greater number of the men have arrived at the appointed spot, for the purpose of directing those who may have gone astray.

§. 50. *Conduct when attacking.*

a) *With Respect to the Place.* If you have approached the enemy in front, you must endeavour, by making a detour, to attack him in the rear. Good guides, to whom a certain part of the booty should be promised, are indispensible on such an occasion.

b) *Of dividing the Enemy's Attention.* When the principal attack is on the enemy's rear or flanks, another should be made by a detachment, sufficiently strong, in front; perhaps you may thus penetrate in some point where you did not expect it; or at any rate, by attacking in different points, you oblige him to divide his force in all quarters.

c) *Precautions necessary in attacking with several Detachments, in different Points, and at the same Time.* When this is your intention, you must, if it is not too hazardous, endeavour to get into the rear with your whole detachment, and not divide it until you have arrived there. Unless this is done, one part generally arrives too soon, or is discovered, and the whole disposition is then ruined. Many surprizes have failed in consequence of the party being thus divided.

d) *Of commencing the Attack.* The party should not fire at the commencement of the attack, particularly if it consists of infantry; every thing must be done with the bayonet. In the attack on Zierenberg, in 1760, the grenadiers did not even load, and the attack succeeded completely. Firing may perhaps throw the party into confusion, and retard its advance, which may give the enemy time to concentrate himself. This rule however has an exception, which is the attack of a post in an open field; if you are able to approach within the distance of 150 paces undiscovered, you then fire a volley and charge immediately: by this you obtain two advantages,---first, the enemy is disconcerted by the unexpected fire, and secondly, the subsequent attack is more impetuous, from the men breaking in with their bayonets. It is to be understood that you do not wait to re-load.

In attacking, the advance must be as rapid as possible, for every thing depends upon the first moment.

When the attack is made by cavalry on open places, they must endeavour to arrive at the alarm posts with the enemy's outguard, and immediately divide into small bodies of four men each, and proceed through the streets, while stronger parties are posted in front of the outlets.

e) *Of the Points to which your Attention should be particularly turned.* One party must proceed directly to the alarm post, and another to the quarters of the commandant, if you have been able previously to inform yourself of the situation of these two places. If you have entered a fortified place by scaling the wall, or by any other means, the first object is to open the gate.

CHAPTER II.

Of the Surprise of an Outguard, or small Detachment.

§. 51. *General Conduct.*

a) *With Respect to the Attack itself.* You must fall at once upon the enemy with your whole force, and not allow him time to recover from his confusion; therefore it is only in particular cases that you should have reserves, or occupy defiles in your rear.

With Infantry. The party must advance secretly to within the distance that musketry will be of effect, if possible, to about 100 paces, and then fire a volley, and rush on with their bayonets.

With Cavalry. Cavalry should approach silently at a walk, until they are discovered, when they must instantly attack with the greatest impetuosity.

With Infantry and Cavalry. As soon as the infantry, who are in front, fire, the cavalry fall upon the enemy's flank.

If you surprise an Outguard with Cavalry, which is supported by Infantry in enclosed Ground. The cavalry only must be pursued, and the infantry avoided, as it would be impossible to effect any thing against them in such a situation.

b) *Of the Enemy being supported.* You must not remain upon the spot, but retire immediately, as the enemy will lose no time in supporting his outguard. It is seldom necessary to detach parties to oppose the enemy's support. The enterprize should be concluded so rapidly, as not to allow his picquets to overtake you; besides which you would not be strong enough to resist them. It is often right, in addition to the detachment destined for the duty of the surprise, to have another in readiness to cover its retreat.

c) Place of Rendezvous. §. 49. d.

d) *SIGNAL.* §. 49. c.

e) Road on Retreating. In retreatiug always take a different road from that by which you advanced, and endeavour to pass through roads where the enemy cannot pursue you.

§. 52. *On the Attack of an Outguard, or a small Detachment in the Rear.*

An instance which was executed strictly according to rule, will give the best idea of undertakings of this nature *. (Pl. I. No. V.)

* This description of the surprise is given by Captain Luderitz, of the Queen's light dragoons.

In the winter of 1759, the French occupied Butzbach, having out-posts in front of the place. The regiment of Luckner hussars and a battalion of grenadiers were stationed at Niedern-Klee opposite Butzbach. A large fire was perceived during the night in a wood that lays immediately in front of Niedern-Klee, and patrols were sent out to discover if it was practicable to attack this post, but were always compelled to retire by the fire of the enemy's infantry; abbatis had likewise been formed which rendered any attempt on that side useless. At length, information was received, in the evening, from a forrester, that the post consisted of 50 infantry and 30 hussars of Chabosch, and they had cleared a square in a spot thickly covered with fir trees, leaving only one opening on the side towards Butzbach: a detour of four miles was consequently necessary to attack them in that quarter. General Luckner charged me with the conduct of this surprise: Lieutenant Niemayer, of the regiment de la Chevalerie, was ordered to join me with 30 grenadiers, at twelve o'clock at night; and having called out the volunteers, between 2 or 300 came forward, from whom I selected 30, and set out with the detachment and the forester, who was provided with a horse. On arriving at the road which leads from Butzbach to the wood, I clearly perceived the infantry in the square standing by their arms round the fire, and the hussars mounted. I immediately halted, and formed the grenadiers in the centre, placing myself with 18 hussars on the right, and a corporal with 12 upon the left, in order that none of the enemy's hussars might escape. Lieutenant Niemayer ordered the grenadiers to make ready, and we thus advanced in as good a line as the ground permitted. When we arrived within 100 paces distance, the enemy became attentive, from hearing the horses moving upon the frozen ground, two hussars rushed forwards, fired their pistols, and retreated on the flanks; upon this our grenadiers fired a volley, which completely disconcerted the enemy: several men were killed, most of their hussars took to flight, and the remainder, including the officer, were made prisoners.

The noise made by the enemy's posts round Butzbach, shewing that they were alarmed, and the garrison of that place being turned out, our detachment retreated immediately with the prisoners: the grenadiers got through the abbatis, which the hussars could not effect. I wished to avoid retiring by the road I had advanced, and making such a detour; and endeavoured to find some open place on the side of the wood, but fell in with another of the enemy's posts, which was concealed, and one of my hussars was killed by their fire. At length I discovered an outlet, but found the country between Butzbach and Niedern-Klee, through which I had to pass, filled with cavalry. In order to ascertain whether they were friends or enemies, I halted in the wood, and directed a corporal, with two hussars, to approach the nearest party; when, from the answer to the challenge, I discovered that they belonged to our own regiment. General Luckner paid me a very handsome compliment, adding, that he conceived, at first, that the surprise had failed, from hearing so much firing, which induced him to advance in front of the village with the hussars and grenadiers.

The enemy's fault was, thinking himself too secure; had he placed a small post in the direction of Butzbach, the surprise would certainly have failed.

You may always be assured, that the enemy will endeavour to revenge himself, and he was very near doing so in the present instance.

In consequence of the cold weather, the hussars and grenadiers were obliged to go the next day into winter-quarters, at Dornholzhausen, about two miles from Niedern-Klee; one troop remaining behind that place as an out-post. The enemy attempted to surprise this troop; but Captain Brinky, who commanded the post, ordered his men to advance behind the hedges of the gardens, several hundred paces distant from the picquet-fire, near which no person remained, except a sutler's wife. The enemy approached from behind the wood at Niedern-Klee, and discharged a volley at the fire, by which, the only person there, the sutler's wife above-mentioned, was killed. The regiment was instantly turned out, and a troop advanced to the support of the post; but the enemy fell back rapidly after their unsuccessful attempt; they were, however, pursued, and seven men taken. They were not yet contented to remain quiet, and were eager to make amends for their faults; in consequence, they attacked

L. Gœns, which was occupied by several rifle companies, but they found them likewise on the alert; notwithstanding which, they succeeded in entering the village, which, indeed, can never be prevented.

§. 53. *Attack in Front on an Outguard, or a Detachment which is at no great Distance.*

If you are in Front of the Enemy, and cannot get into his Rear. This enterprise is very dangerous, unless favored by bad weather and dark nights, relaxed vigilance on the part of the enemy, or perfect information of his position, &c.

You must move at night from your station, and approach the enemy silently. If your party consists of infantry, a few non-commissioned officers should be sent forward, who must creep as close to the post as possible, while you halt in the rear with the remainder. These non-commissioned officers must secret themselves, and when the patrols have passed by, the sentries have been relieved, and they have acquired a knowledge of the post; they will give notice of it to their commanding officer. The detachment, when it consists of infantry, as also the non-commissioned officers, who are sent out for information, should lay down and conceal themselves in pits or high corn by the road side.

Every thing else must be executed as pointed out in §. 51.

The following instance will shew, how easily an enterprise of this nature may be executed when the circumstances on which it depends are favorable.

One of the blockading detachments of the fortress of Munster, in October 1759, consisting of one battalion and a half, and one squadron from the allied army, was stationed upon the heath of Driburg, about 4000 paces distant from the nearest works.

This detachment had been reconnoitred by M. de Boiscleran, second in command of Munster, who was ordered to attack it on the night between the 17th and 18th---725 men were appropriated for this service; these he formed in three divisions, which, keeping near to each other, approached in front silently and slowly towards the Hanoverian detachment, as soon as they were discovered, they rushed precipitately into the camp of one of the battalions, so that the men had not time to get out of their tents. The whole detachment was routed; 1 gun, 19 prisoners and 40 horses were taken, and 4 officers and 35 men wounded.

§. 54. *Attack of an Outguard or a Detachment situated at a considerable Distance.*

An example (which is supposed) will best shew the manner of taking advantage of the local or other circumstances, in such cases.

Suppose that Neustadt is occupied by the enemy (Pl. 1. No. IV.) and an outguard is stationed in front of that place towards Empede, which, having the morass on the left, and the river Leine on the right, cannot be turned. A detachment of 60 cavalry is sent from Hudemuhle and Ahlen to reconnoitre the country as far as Neustadt. On arriving at Mandelslohe, the detachment receives information that Nienburg and Neustadt only are occupied by the enemy, and that an outguard is posted in front of the latter place, upon the road to Mandelslohe. In order to ascertain this more certainly, the detachment approaches, by a night march towards Empede, and places itself in ambuscade in the neighbouring forest during the following day; it is conducted from Mandelslohe by a guide, who is a native of Mariensee, and is well acquainted with that part of the country. The outguard cannot be seen from the forest near Empede, as soon, therefore, as it is dark, some persons must be procured from this village (see §. 37. c.) and interrogated. From these men you learn that the enemy patrols during the night no further than the mill near Empede: you therefore detach a non-commissioned officer and one man to that spot, who are to remain concealed in the reeds by the side of the ditch, and observe the enemy's patrols, and to acquaint you as soon as one has passed: upon this you cross the bridge, which should be covered with straw, to prevent the outguard from hearing the horses: the trumpeter and one man must remain here to give the signal after the attack to such men as may miss their way.

The detachment does not march upon the road, but paralel to it across the field, and as soon as it is discovered, attacks the enemy in three divisions. The two flank divisions advance a short distance before the centre, and extend in front, each having a non-commissioned officer and four men as a reserve in its rear. If you cannot reach the outguard before the men have mounted their horses, or if you cannot get into its rear, you must notwithstanding pursue instantly, and as rapidly as possible; on which every thing depends. The prisoners must be taken directly to the bridge.

It would be dangerous to approach within musket shot of Neustadt.

The retreat must be effected by way of Mariensee, and several planks must be pulled up from the three bridges which you pass on this route, particularly if you understand that the enemy have cavalry in the neighbourhood of Neustadt.

§. 55. *Instance of a Surprise of an Outguard.*

During the Bavarian war of the succession, in August 1778, the army of Prince Henry was posted near Nimes, (Pl. II. No. II.) and that of Laudon behind the Iser: Hoflitz, Neudorf, Plauschnitz, &c. were occupied by outposts; there was likewise an advanced post at Hunerwasser, and a party of 30 infantry was placed in the monastery upon the hill of Posig. An outguard of 30 infantry was stationed near Unter-Wocken, at the corner of the forest (*a*) in front of which, towards Wocken, 30 hussars were posted during the day, at (*b*), who at night retired behind the infantry. For the purpose of surprising this outguard, and also of alarming Plauschnitz and Neudorf, a party consisting of 50 hussars and 80 dragoons was detached from Munchengrætz; and 100 infantry advanced as far as Jablonitz, to cover its retreat through the defiles.

The detachment set out from Munchengrætz at midnight,

and proceeded by way of Witzmanov and Jablonitz, until it got into the rear of the post at Hunerwasser; from thence it marched through the forest as far as Wocken; and debouchecd at day-break, about 100 paces distant from the outguard.

The outguard, on being instantly attacked by the hussars, fell back upon the post of infantry, who fired, but without effect. The cavalry were pursued through the forest as far as Plausschnitz, where a company of the regiment of Hord, being under arms, nothing could be effected, and some loss was sustained. The 80 dragoons had remained as a support in front of the wood near Nieder-Wocken. As soon as the alarm became general, the Prussian hussars turned out; they however were unable to overtake the party, which retired rapidly, but formed twice upon the plain to allow the dragoons time to retreat.

The detachment of infantry prevented the Prussians pursuing the cavalry through the defiles of Jablonitz.

Remark.

1. The infantry in the village of Plauschnitz should not have been engaged.

If it was supposed that the enemy's cavalry might have been attacked with success, the first moment should have been seized for that purpose. But 50 dragoons ought to have been drawn from the support; for it was evident that *nothing* could be effected against the village by the hussars alone; *even if there had been no infantry.*

2. The detachment arrived too late near Wocken, they should have set out sooner; and if it was intended to make any attempt during the day, they shonld have waited until the horses of the outguard were fed; but the men were mounted, when the detachment quitted the forest, on which account they easily escaped. Perhaps the commanding officer was afraid of being discovered if he remained in the forest.

CHAPTER III.

Surprise of an Enemy's Quarters, or of a considerable Post or Place which is not fortified.

§. 56.

Having obtained the requisite information as far as possible, through patrols, peasants, travellers, spies or by any other means (§. 43.) and finding that some of the circumstances exist that have been stated in §. 44, you make your *disposition* before marching.

a) *Disposition*, §. 49.

b) *Division.* The detachment should be formed in as many sub-divisions as there are outlets to the village; one must be appointed on which the others are to rally after the attack, and another to oppose such parties as may advance to the enemy's assistance.

For instance, if a village has three outlets, and you attempt the surprise of it with 200 cavalry; 50 men should be appointed for the attack of each outlet, 30 for the division which is to oppose the succours arriving to the enemy, and 20 for the reserve on which you propose to rally after the attack.

If the detachments consists of infantry and cavalry, and you propose to surprise an open village, situated near a forest; the infantry formed in two or three divisions, will advance on the side of the forest, while the cavalry approach on the other.

If you make an attack, with infantry and cavalry, upon a country town; and have not positive information, whether it is perfectly open or barricaded, each division should consist of both cavalry and infantry, and the latter will remove with their axes any impediments that may exist.

If you are obliged, on your retreat, to cross considerable plains terminating in defiles; it will be frequently necessary to leave the whole of your infantry in reserve at the defiles; particularly when you apprehend that the enemy will be strongly supported. In an intersected country the reserve must always consist of infantry; and is often stronger than the whole of the attacking party: for instance, when the post you intend to surprise is not strong, but may be powerfully and quickly supported. This was the case in the attack on Zierenberg under the Crown Prince in 1760. Several squadrons were pushed on beyond Zierenberg to meet the enemy.

c) *Disposition for the Attack previous to Marching.* The commanding officer should communicate to the commandants of divisions as far as can previously be done the disposition for the attack. He should describe to them the situation of the place, and the roads he intends to move by.

The divisions which are to make the attack, halt in front of the outlets of the village; and that which is to oppose the enemy's succours, proceeds along the road by which it is probable they will arrive. The reserve division on which you are to rally after the attack, remains on that side of the village by which you purpose to retreat.

d) *Signals, which are to be previously fixed.* The signal for retreat, to be given by the division appointed to oppose the enemy's succours, in case they should force their way, should be the beating or sounding a certain march; which should also be repeated by the division in reserve, for the purpose of indicating the rallying point. In addition to which, the sign of recognizance should be communicated to every individual. (See §. 49. c.)

e) *Of the March.* The march should be regulated according to the rules laid down for secret marches, §. 36. &c. The formation of the detachment for the attack should likewise be attended to on the march. The division appointed to oppose the ennemy's succours should be preceded by that which is to form the attack, and the party destined as a reserve should compose the rear guard.

f) *Conduct to be observed by the Divisions.*

Cavalry. Each of the three divisions, composing the main attack, should be formed in three sections, 1 or 2 of which, as soon as the division is arrived in front of the outlet of the village, will disperse through the streets and yards, in small parties of 4 men each, in order that the enemy's men may be taken singly.

Any persons seizing horses must take them immediately to the section, which remains formed at the outlet; and where the whole are to assemble.

The division ordered to oppose the enemy's succours, must, if possible, occupy any defiles which it is necessary for them to pass; or, at all events, several small parties should be detached from the main division upon the different roads, by which it is probable they will advance, while the division itself remains in the centre in the rear of the defile.

The reserve should send from time to time to the parties stationed at the outlets, for the purpose of taking charge of the prisoners.

Infantry. The divisions, when composed of infantry, do not disperse; one section remains at the outlets while the other division enters the streets, (the men keeping their ranks,) and sends small parties in the by-streets and houses, where the enemy might be concealed.

The men should be furnished with axes, for the purpose of forcing open the gates and doors.

Firing creates delay and confusion, and in the night your own men are easily mistaken for the enemy; no firing therefore should be allowed; for which reason it is better that the men should not load, for if they do, it will be difficult to prevent their firing.

Infantry and Cavalry. When the division, intended to attack an open country town, consists of both cavalry and infantry, the latter should be in front at night. They should also be provided with axes and shovels, to remove any obstacles, concealed barricades, gates, &c. in order to enable the cavalry to act, which they afterwards support, and attack any body of the enemy that assembles together; but if the cavalry can rush in upon the first alarm, they will generally succeed.

g) *Consideration of the Measures necessary for a Retreat.* As enterprises of this nature are frequently betrayed or suspected by the enemy, and as in that case the detachment is in great danger of being attacked on the march and cut off, you must, in the very first instance, fix upon a different road than that by which you advance, and which should be pointed out to the commandants of the different divisions, in order that every one may know, under any circumstances, how to provide for his own safety. If there are any defiles in your rear, they must be occupied as pointed out in §. 47.

h) *Respecting the Enemy's Alarm Post and the Quarters of the Commandant.* See §. 50.

§. 57. *Examples of Surprises of Quarters and Posts.*

As the surprise of a post depends upon particular circumstances, the following examples are necessary to shew how circumstances even the most trifling are to be turned to advantage.

a) ***Surprise of a Cavalry Post by Cavalry.*** It is the negligence of the enemy in placing his outguards and sending out his patrols, which generally affords the opportunity of surprising him: the most correct information of these circumstances must however be procured; in our own country this is easily done, by means of peasants, who are acquainted with the places occupied by the enemy, or who visit them either upon their own concerns or by our own instigation.

When the allied army, in February 1758, moved from the neighbourhood of Luneburg towards the Aller; (Pl. I. No. III.) Polerezky's regiment of French hussars was stationed in the villages of Bothmer, Schwarmstedt, &c. in the district of Essel, (Pl. I. No. IV.) Information having been received that the allies were marching upon Hudemuhle and Eickeloh, and approaching the right of the Leine, part of the French hussars proceeded towards Stœckendrebber, leaving a detachment at the bridge of Bothmer, which secured that point: another strong party directed its march towards Neustadt, and remained on the right bank of the Leine: this river had overflowed, and as far as Neustadt was impassable, except by the bridge of Bothmer.

In this situation, the whole of the allied army being on the other side of the Aller, and Nienburg (Pl. I. No. III.) in possession of the French, the hussars considered themselves in perfect security at Stœckendrebber; but, in order to receive timely notice, if the allies should pass the river and advance, an officer was directed, on the 23d of February, to proceed with 20 cavalry, by the way of Hœren, towards the Aller (Pl. I. No. IV.) for the purpose of gaining information; this party arrived near Hœren, when a detachment of Prussian hussars, after crossing the river near Ahlen in boats, reached Gilten.

At Gilten the Prussians were informed that the bridge at

Bothmer was occupied, and that the main body of French hussars was at Stœckendrebber; a peasant of Gilten, who had just returned from that place, also acquainted them that the French were about 300 strong, and had only a village guard and a vedette in the field at *a*. (Pl. I No. I.) The peasant was promised the best horse that should be taken, and a further remuneration, if he would conduct the Prussians by some unfrequented and circuitous road to Stœckendrebber.

The detachment commenced its march from Gilten, in perfect silence, about ten o'clock at night, and proceeding by Neuhagen (Pl. I. N. IV.) towards Stœckendrebber, got iuto the enemy's rear.

On arriving at the wood *b* (Pl. I. No. I.) the Prussians divided themselves into two parties. One proceeded towards the vedette at *a*, in order to enter the village by the rear; the other advanced in front and halted until they heard a shot fired by the former party. The vedette was put to death before he could fire; a non-commissioned officer, however, who had seen this party, alarmed the colonel and several men, who were quartered at *f*, so that the principal part of the French had saddled their horses before the Prussians entered the village. The latter dispersed themselves on every side of the village to collect booty: they took aboat 150 horses and the regimental chest.

Polerezky, the commander, was severely wounded and died in consequence. One officer, the vedette, and another private only were killed on the spot. The greater part escaped, owing to the Prussians dispersing and not occupying the outlets. It is extremely difficult in such cases to restrain the eagerness of the soldiers for plunder.

§. 58. *Surprise of a Corps of Infantry in Quarters, by Infantry and Cavalry.*

Thadden's regiment of Prussian infantry was surprised, in 1759, when it was cantoned in the village of Dittersbach, (Pl. III. No. II.) five squadrons of hussars lay, at the same time, at Pfaffendorf, and five squadrons of dragoons at Schreibendorf; the post of Altweissbach was occupied by 1 captain, 3 subalterns, 6 non-commissioned officers, and 100 cavalry, dragoons and hussars; who detached 1 officer and 30 men to Neuweissbach.

The regiment of Thadden at Dittersbach had one non-commissioned officer and 12 privates stationed upon the road to Michelsdorf and Patzelsdorf at *g*: in addition to which, there were redoubts with guards at *d*. and *c*. and guards at *a*. *e*. and *b*. The Austrians were beyond A. behind the mountains, and extended as far as Michelsdorf.

General Wurmser, who commanded the Austrians, having obtained correct information of the Prussian quarters, resolved to cross the mountains, with four battalions of croats and four of hussars, and surprise Thadden's regiment at night.

For this purpose the infantry were formed in two columns and passed over the mountains, while the cavalry descended on the side of Michelsdorf. The attack on the part of the village between *e*. and *c*. and the redoubt *d*; was to be made by the first column of infantry; and that on the remainder of the village and the redoubt *c*. by the second column. The disposition of the first column was as follows: two companies were to enter the village for the purpose of making prisoners; two were to hold the redoubt *d* in check, but not to attack it, as it was considered too strong, and it was wished to avoid sacrificing many men; the two remaining companies were to direct their attack on *e*, where the colonel was quartered, and a picket was stationed. The disposition for the first battalion of the first column was carried completely into effect, but the other battalions did not come up, as had been directed, to the support of the party, which should have seized the picket at *e*. The column, which was to have attacked the other part of the village, in the manner above-mentioned, having missed its way, also arrived too late.

Eight regimental colours and 65 men were however taken, and the colonel, his adjutant, and about 20 men were killed and wounded.

Observations.

1st. This enterprize proves that it is disadvantageous to advance in more than one column, as such an arrangement generally creates confusion, and increases the danger of discovery, at least of one of the divisions; this was the case with the cavalry in the present instance: a hussar and a dragoon, who were in search of a deserter, fell in with this column, the dragoon escaped, and gave notice of the advance of the enemy.

2d. From this example, the following practical rule may also be deduced, namely, that the surprise of strong places, such as forts, redoubts, &c. should never be attempted; but that attacks of this nature should be directed only against posts which are not in a proper state of defence. The works in the present instance were of little importance.

3d. The attack was not made with sufficient rapidity. The distance from the upper end of the village *c.* to *e.* is only 800 paces, and had the troops advanced to *e*, in the first instance, and then divided themselves into small bodies, it is probable the issue would have been more successful.

Surprise of a Post of Cavalry by Day. In this case every thing must be executed with rapidity. The campaign of 1759 will afford an example:—

General Luckner was detached from the camp at Marburg, on the 22d of October, 1759, for the purpose of driving the enemy from the neighbourhood of Dillenburg. On approaching Weyer and Munster, he learned that there were 400 cavalry in the village of Niederbrechten, and 150 in Oberbrechten; and he determined to endeavour to surprise the former, as he found he could get into the rear of the village; and enter in that direction; he was discovered, however, by a patrol, consisting only of three men; he immediately ordered the patrol to be pursued at full speed, and entered Niederbrechten with it; a part only of the enemy had time to mount, and the result was, that 70 men, 90 horses and 100 forage waggons were taken, and 43 men and 9 horses killed.

§. 59. *Of the Surprise of several Posts and Quarters.*

An example will here supply the best instruction.

When the allied army, in 1759, was obliged to retire upon Minden, and the French made themselves masters of the rivers

Weser and Werra, having detachments in Witzenhausen, Hedemunden, Munden, Dransfeld, Hemeln, &c. Lieutenant-Colonel von Freytag was stationed with his corps of Yagers in the wood of Sollingen, to cover, as far as possible, the district of Gottingen, &c. He conceived that he should perform the service required of him better by undertaking offensive operations, than by maintaining a position merely defensive, which in such cases is seldom of much avail against an enemy constantly pressing forward; and he therefore determined to penetrate within the enemy's quarters. The object here was to seize some of the enemy's detachments, to pursue the others closely, and then take advantage of their consternation in the first moment.

The volunteers of Elsass were stationed in the village of Hemeln, which is situated about four miles from Munden, between the Weser and the forest of Sollingen. This village was surrounded by wood, and its security depended upon the patrols which were sent out. Colonel von Freytag having seized one of these patrols, (in the manner described in §. 14), at Uslar proceeded immediately in the direction of Munden, until he arrived in rear of Hemeln, upon which place he then directed his course. He marched in the night between the 4th and 5th, and at daybreak was still two miles distant from Hemeln. In order to prevent the arrival of succours, and to intercept the fugitives, he caused the roads towards Munden and Burschfeld, in which places the enemy had strong detachments, to be occupied; and with the remainder of his corps attacked Hemeln in two points. The French endeavoured to retreat towards Munden, but finding that road occupied, they directed their flight towards Burschfeld, but with no better success: as a last resort, several tried to cross the Weser in boats, but most of them perished in the attempt. As soon as this blow was struck, Colonel Freytag moved against Burschfeld. The village was surrounded, and a detachment of the volunteers of Elsass, which lay there, was taken. The Colonel, Beyerle, 27 officers, 30 non-commissioned officers and 186 privates were made prisoners in Hemeln and Burschfeld; and the greasest part of the remainder either killed or wounded.

Lieutenant-Colonel Freytag now marched towards Munden, but learning on his way that a reinforcement had arrived at that place from Cassel, and that a strong detachment was encamped in front of it, he altered his plan, and proceeded that day as far as Juhnde; and as soon as it was dark, he caused the enemy's outposts, which extended from Munden towards Niedernscheeden, to be attacked by Captain Bulow, with a view to create a belief that his design was against the former place: at the same time, a small detachment of the enemy, which was employed in pressing waggons, was seized at Dransfeld. At day-break, the Lieutenant-Colonel directed his march to Witzenhausen, and the enemy remained in ignorance of his movements until he appeared before that place. A company was stationed at Hedemunden, to secure the right flank, and to prevent any succours arriving to the enemy's assistance from Cassel and Munden. The attack upon Witzenhausen was executed in the following manner: one company of yagers was sent across the Werra, to attack the place in the rear, while two other parties approached it in different quarters. The town, having been summoned and no answer given, was stormed; the gate was forced instantly, and the principal part of the garrison either killed or wounded; the remainder, consisting of one Captain, two Lieutenants, and 69 privates, partly cavalry and partly infantry, were made prisoners.

From the success of this enterprize, the following rules may be laid down for imitation in similar cases:

1st. It is necessary that the march should be conducted with the profoundest secresy. If this had not been done in the above instance, the patrol at Uslar would not have been seized; the corps had until then marched through the forest in the greatest silence, and remained undiscovered even by the inhabitants of the villages near which it passed.

2d. The first and principal step towards ensuring the success of a surprise, is the seizing a patrol, although it is in itself a trifle. In the above case the French relied upon the patrol, and when that was expected to return the enemy appeared.

3d. If it is possible to proceed, during the night, so far as to get completely into the enemy's rear, a surprise may succeed even by day, as was the case here; even if the enemy dicovers you, it is too late, as he is already cut off. But if you wish to succeed, you must, like Colonel Freytag, make a long detour.

4th. In case the enemy should discover you, it is necessary that all the roads, by which he might retreat, should be pre-occupied, or some part may escape. If this precaution had not been taken, very few prisoners would have been made in the above enterprize.

5th. Detachments must be stationed to oppose the enemy's succours. This general rule, already mentioned, is repeated here, and the more so, as the enemy's troops cantoned in the neighbourhood will immediately turn out on any alarm.

6th. If the design has succeeded, advantage should be taken of the spirit and confidence which naturally animate the troops, and of the alarm and confusion of the enemy, to attack other posts, particularly those at some distance, who will least expect it. As was the case with respect to Witzenhausen.

7th. In order to prevent the troops in other quarters from getting into your rear, or supporting the part which you purpose to attack, they must themselves be threatened, and by this means kept in a state of uncertainty. This was done by the attack upon Niedernscheeden.

Thus a well-executed attack furnishes many rules, which will not be discovered by those who do not reflect upon the subject.

CHAPTER IV.

Surprise of a Post stationed in a Castle, Town, &c. by a small Detachment, without scaling the Walls, or destroying the Gates with Artillery.

§. 60.

In a surprise of this nature, the great object is to get into the place undiscovered.

You may approach during the night, and remain concealed in front of the gate, until it is opened, and then rush in. There are several instances of gates having been taken possession of in the day time, by a party in disguise, or concealed in a waggon of hay. Mr. von Wintzingerode did not succeed however in his attempt to surprise Sababurg, in 1760, in this manner. The French were more successful when they adopted it against a post on the river Hufe, in 1760. You may sometimes contrive to steal into a place quietly during the day, before the gates are shut.

Some examples will serve to elucidate the subject.

a) The allied army in November, 1759, was posted with its right towards Giesen, and its left towards Marburg, being separated from the French by the river Lahn, and occupying Herborn, Dillenburg, &c. on its right flank.

Lieutenant Du Près, of the royal legion, was ordered by Count Chabot, who commanded in the neighbourhood of Giesen, to make an incursion into the country occupied by the right wing of the allied army. For this purpose, he crossed the river

Lahn with 50 dragoons, and having received intelligence that 60 hussars and 20 infantry were stationed at Herborn, he determined, if fortune favoured him, to surprise them, and, with this view, reached the place about two o'clock on the morning of the 9th. Having received information from a man, whom he had sent forward in disguise, respecting the situation of the post, he approached the gate, and formed his dragoons in the best order circumstances would admit, close in front of an old building; he then caused twelve men to dismount, and advance silently towards the gate, with orders to make themselves masters of it as soon as it should be opened. These men got within twenty yards of the gate, and laid themselves down behind a rising ground; they heard the sentries relieved and the soldiers talking, and they allowed a patrol, which was sent ont, to proceed quietly, but when it had passed the main body, it was attacked in the rear and taken.

The prisoners were required to cause the gate to be opened, and to conduct the detachment into the town A Hanoverian rifleman was willing rather to sacrifice his life than perform this, but a Prussian hussar consented; he therefore pretended that the patrol was returning, and the gate was opened by his desire; the sentry was put to death, and the greater part of the garrison killed or taken, together with 27 horses.

b) The gates are in general open during fogs, and if you can approach the place undiscovered, you may perhaps push in before the guard has time to shut the gate, or raise the draw-bridge.

When the Marquis of Armentier moved from Dorsten to Bocken, in 1759, the French left a post of 130 men in the former place.

General Imhoff, who was besieging Munster, detached Major Bulow, with a strong party of infantry and cavalry, for the purpose of seizing this post. He arrived, after a forced march, during a fog, at a thicket in front of the bridge over the river Lippe and from thence directed some grenadiers of the corps of Scheiter to approach the sentries silently, and put them to death without firing, while he followed at a distance with the remainder of the detachment.

The first sentry, in front of the bridge, was seized without firing: but a coporal, who was on guard with four men in an old building, discovered the party and retreated in safety; he was however pursued so rapidly, that there was not time to shut the gate. The alarm was instantly given, but notwithstanding four officers and 80 men were taken; the remainder fled into the thicket through the gate towards Wesel, which place they reached the same evening.

c) A surprise often succeeds during the day, from the enemy not expecting it at that time. In this case, it is necessary to approach rapidly and secretly, or to pass undiscovered by some guard, upon which the enemy relies.

When a corps of the allied army in 1760, was stationed in the country of Einbeck, and the French had some detached corps on the Eichsfeldt, a detachment of the former, consisting of one officer and 24 cavalry, was posted at Northeim for the purpose of drawing forage from that neighbourhood. The French having dislodged this party and occupied the place with 600 infantry and some cavalry, under the command of Colonel de Meer, by which they were enabled to forage the whole of that country, Major Bulow was directed to attempt the recovery of it, having under his order, for that purpose, half a brigade of the rifle corps of Freytag, 80 of the Luckner hussars and 150 infantry.

Some representation of this place is given in Pl. V, No. 1. The French had guards at the gate *a. c.* and *e.* and an outguard at F. in front of the bridge over the river Ruhme, which is about 1200 paces from the town. The guard F. was posted in the inn called Rucking, on this side of the river, and had only a double sentry immediately in front of the bridge, on the road to Einbeck. The Ruhme is not fordable every where, but may be passed at many places in the summer.

Major Bulow had intelligence of the situation of the guard at F. and he was acquainted with a ford by which he could pass the river at A without being perceived by it.

This was so successfully executed, that he arrived in front of the gate *e*, by the way of A before the French were aware of his approach. The sentries had just time to shut the gate; it was however forced, and to encrease the alarm, a small gun was fired in the street leading to the mansion-house *b*. The French were closely pursued, but rallied in the triangular church-yard, in front of the mansion-house near *a*. Part of the cavalry had, at the commencement of the affair, been sent round the town to B, where it was posted to prevent the escape of the French.

The French had now no alternative but to surrender, or force a passage sword in hand: they preferred the latter, and formed a square near *a*. in which manner they withdrew into the mountains by the way of C. The mounted riflemen hung upon them, and killed several officers in the ranks, but were not sufficiently strong to accomplish more. Had horse artillery been then in use, and a few guns attached to this party, the square might have been thrown into confusion. But notwithstanding 300 infantry and 32 dragoons were made prisoners.

d) Another instance of a surprise, which took place during the campaign of 1760, will shew how apparent impossibilities may be executed with cavalry, by rapidity of movement.

When a corps was stationed, in the month of May, near Kirchhayn under General Imhoff; General Luckner was detached with his own regiment of hussars and a party of grenadiers, for the purpose of intercepting a convoy between Giesen and Butzbach. Having proceeded as far as L. Gœns, and waited for the convoy in vain, he at length determined, as he could not succeed in his first object, to attempt the surprise of Butzbach with his hussars; the place was occupied by both cavalry and infantry, and although it was not fortified, it was impossible for cavalry to enter it, if the gates are closed in time.

General Luckner, in his report to Duke Ferdinand, relates the affair as follows:

" I gave orders to push through the gate at speed; and if it " was not possible to effect this, to pass rapidly round and enter on " the side of Friedberg; or if the infantry retreated in that " direction, to charge them impetuously. I also directed my infan- " try to cannonade the gate with a small four-pounder, and then " to rush in with the bayonet. A misfortune occurred at the " commencement; eleven men only were secured out of a " patrol consisting of a cornet and 12, the cornet and one man " escaped; I immediately gave orders to pursue and enter the

" town at all hazards. This was done, and the guards, which were " the only troops to be seen, were in an instant put to the sword. " The soldiers demanded where the enemy was, and being an- " swered that they had retired by the Friedberg gate, rushed out " and made 20 dragoons prisoners; the remainder jumped off their " horses and escaped through the hedges. We secured also 20 " men of a picquet of infantry which we fell in with. Having per- " ceived some cavalry on the road to Friedberg, I directed Cap- " tains Breyman and Bruns to pursue them with their troops; " they succeeded in taking several dragoons, together with 50 " forage waggons. I also sent Brinky into the forest in search of " the enemy, with order to fire and shout aloud, if he discovered " them, when I would come to his assistance; but I learned from " a few that were found, that Waldemer had dispersed his infan- " try in small parties of from 10 to 20 men each."

The above enterprize, and that also of Luckner, mentioned in §. 58. shew how advantage should be taken of the first moment. If, in these cases, the attack had not been conducted with impetuosity, and every thing executed with rapidity, nothing would have been done, and no prisoners would have been taken. It may be observed also from this, that well mounted cavalry may be of the greatest service in a surprise.

CHAPTER V.

Of an Attack with the intent to surprise a Post situated beyond a River which is not fordable.

§. 61.

An attack of this nature has the best chance of success, when you are at a considerable distance from a river, and that it is very broad; as the enemy not expecting the attack, and neglecting the necessary precautions, is disconcerted when he discovers your approach.

a) When Gen. Wangenheim in May 1758, was directed to draw the enemy's attention towards Duisburg; with a view to prevent his intention of crossing the Rhine on the Dutch frontiers from being suspected, he ordered Capt. Scheiter, with his new levied light troops, to undertake something against the French, who were posted on the opposite bank of the river.

For this purpose, Capt. Scheiter having already reconnoitred the banks of the river; on the 28th of May procured, by means of the country people, 13 boats from the river Roer; and made dispositions for arriving before day-break on the bank of the Rhine, at the confluence of that river with the Roer.

The day however broke, as it frequently happens in such cases, before the preparations for attack were completed. The French perceived the vessels upon the Rhine. In the first boat were 40 grenadiers, who, upon being challenged, replied in French that they were friends; the enemy however soon discovered their intention, and after a volley of musquetry, opened a fire from six guns in a battery directly opposite to the boat; the first shot struck the mast; the grenadiers, although they had never before seen an enemy, without regarding this, took to their oars, and on getting into shallow water, without hesitation jumped overboard, advanced rapidly to the battery, and closed with their bayonets upon the French, who having till then continued firing, threw down their arms and fled towards Homburg.

The grenadiers, encouraged by this first success, without waiting for their comrades, who were crossing in the other boats, pursued the enemy without respite to Homburg, putting such to the sword as they came up with. In the mean time, being joined by about 160 men, they divided themselves through the streets, and the French not having time given them to rally, the two battalions stationed at this place, were defeated by less than 200 men. Whilst this was taking place, Capt. Scheiter, with about 20 or 30 dragroons, (the whole he could bring over) advanced to their support, and by his able dispositions, kept a whole regiment of cavalry in check a considerable time; during which, the grenadiers conveyed on board the boats the regimental chest, a quantity of cloth, the colours which they found concealed in a chimney, and a good deal of other booty—they also carried off five of the guns, the sixth which could not be got away, from the enemy pressing upon them in all directions, was sunk. The whole affair was concluded in about two hours and a half.

b) When, after the campaign of 1759, the allied and French armies were in winter quarters; the former in the counties of Munster and Paterborn, &c. and the latter behind the Rhine, &c. Major Scheiter crossed that river near Dulmen, and surprised the small town of Ordingen, which is situated upon the left bank of that river. Major Scheiter had three difficulties to overcome. 1st. To perform a march in winter, so rapidly as to prevent the enemy from receiving information of his advance. 2. To cross the Rhine. 3. After having crossed, to complete his enterprize before his retreat to the river could be cut off.

He got the better of the first difficulty, by procuring waggons for his men, and did not allow them to march on foot until they arrived near their destination. He crossed the river in a boat, which he took from the river Roer near Ketwich, and had conveyed to the Rhine upon a waggon with 12 horses.

In order to avoid being attacked on the first alarm, by all the troops quartered in that vicinity; he made choice of the night preceding new year's day, that the noise of the attack might not, in the first instance, be distinguished from the beating of drums and firing, which is customary at that time. He succeeded completely, forced his way into the town, which was occupied by 150 men of the regiment of Jennert, set fire to the magazine (at that time however not very considerable) and took 40 prisoners and a great deal of baggage.

He might have made more prisoners, but he was not particularly careful to prevent their escape, as he turned his attention solely to the magazine and the security of his retreat: for this purpose he detached parties in the first instance, to the right and left to oppose such of the enemy's troops as might attempt to advance, and he left another party with the boats.

SECTION IV.

Conduct of Outguards and Picquets of Infantry and Cavalry.

CHAPTER I.

*General View of the Arrangements necessary for the Security of a Camp *.*

§. 62. *Usual Dispositions for the Security of a Camp **.*

a) *Outguards.* In order to secure an army from surprise, outguards are posted at from 2000 to 6000 paces distance from it. These outguards, in an open level country, are composed of cavalry, and in woods, villages and in intersected ground, of infantry; they post sentries and vedettes in their front, that they may not be surprised. The sentries and vedettes, with the guards, should form a

* In order to judge correctly of the proper method of posting outguards, it is necessary to take into consideration the connection in which the guards and posts stand with respect to each other and to the whole. Therefore an officer should have a general idea of the arrangements which should be made to secure an army from surprise, either when encamped or in winter quarters. This has induced the insertion of the present chapter, for officers detached with outguards, &c.

** The dispositions of security here pointed out are the same as were adopted in the allied army during the seven years' war.

connected line, through which an enemy cannot pass undiscovered. This is called the chain of outposts.

These guards are posted only towards the enemy, or in such directions as he can advance. When there are posts of considerable consequence in the country, such as strong castles, country towns, large villages, or particular spots of ground, the possession of which is important, they will be occupied by separate corps which will have their own guards.

There are not so many outguards in general posted in day as during the night; for in most countries, a small number will be able to give fully sufficient notice of the enemy's approach. During the seven years' war, an outguard of 50 cavalry was stationed on each flank of the allied army, when the enemy was at a considerable distance.

b) *Picquets.* It is generally in the night that an army may apprehend a surprise, as it is impossible to observe thee nemy's movements if he is at any distance. For this reason every army has night-guards or picquets; that is to say, there is a certain part of the army constantly in readiness to meet the enemy. During the day they serve as supports to the outguards, if necessary; but at night they frequently march out and occupy the country through which the enemy might approach the camp, and they then become real outguards. They often remain at their stations in the day-time also, when the advance of the enemy's light troops, the apprehension of an attack, an intersected country, or a foggy day may render it necessary.

When a picquet marches out either day or night, it is called an *out-lying picquet*, and is immediately replaced in the camp by another, which has been previously ordered, and until then, denominated the *reserve picquet.*

In the seven years' war the picquets of the second line of cavalry and infantry, occupied the roads in the rear, in order to prevent desertion as well as to secure the army from surprise in that quarter during the night.

c) *Colour and Standard Guards.* In addition to the above arrangement, every battalion of infantry has its colour guard, and every regiment of cavalry its standard guard in their own front, the former at 300 paces distant, the latter quite close. The colour and standard guards of the second line are placed, the former at 300 paces distant, and the latter close in the rear, and facing to the rear. The guns are with the colour guards, and a chain of three double sentries is formed in front of them. The standard guards form the chain close in front of the regiments. Besides these, each regiment has its *rear guard,* which, when there is no second line, form a chain of posts to the rear, and are then reinforced during the night.

All outguards and picquets are under the orders of a field-officer, who takes the command when they are called out, and the whole are commonly under the direction of the general of the day.

§. 63. *Instances of the Dispositions made for the Security of Camps.*

a) *When close to the Enemy.* The outguards are stronger than otherwise. They are supported by the picquets, and by squadrons or battalions, which are posted immediately in their rear, according to the nature of the ground, whether open or enclosed. In this case neither the outguards nor the picquets must be too weak, or an enemy either reconnoitring or advancing to attack, would route and disperse them before they could be supported from the camp.

§. 64. *First Instance.*

When the French army, in 1759, was posted with its right wing at Minden, and its left behind the marsh near Hahlen, occupying also Hille and Hahlen; the position of the allied army was, in the first instance, near Petershagen, having the corps of Wangenheim, which was behind Thonhausen, in its front. (See the position in the plan Pl. III.)

The out-posts were placed as follows: the cavalry before Thonhausen at I. The main guard close in front of the village, having a detached post to the right, near the cross road, and another in its front near the Weser,—the vedettes are marked on the plan, the double ones with a crotchet at the top. During the night, the detached parties, and a troop, which was advanced on the road to Minden, were obliged to remain constantly mounted. In the rear of the outguards, 2 squadrons of hussars were encamped at II. to support them in case of necessity. These hussars, besides a guard and night picquet, had an outguard at III. which furnished the vedettes as pointed out. The picquets were stationed near Holthausen, the cavalry on the left at *x*. and the infantry on the right at *y*. with their vedettes at about 300 paces distant around them.

A detachment of 300 infantry and 50 cavalry was placed at Fredewald, which posted its infantry outguards as shewn on the plan; the cavalry outguard was at Dutzkamp, having its vedettes on the heights marked out; at first, when the French hussars were at Hille, this outguard was drawn back to IV. during the night. In addition to this, a battalion of grenadiers occupied Thonhausen, for the support of the outguards.

The main body of the army, after leaving its camp near Petershagen, took a new position to the left of the village Hille at B B. The corps of Wangenheim remained behind Kuttenhausen.

The following was then the disposition of the outposts. The picquets of the British infantry were near Hartum at *b*; the Hanoverian near Sud-Hemeren at *b*; the Hessians in the wood between Hartum and Holthausen near *b*; the Brunswick near Stemmeren at *b*. The picquets of Wangenheim's corps were in front of Kuttenhausen. The cavalry picquets were posted in front of Holthausen, at *b*, *c*, and a party was detached towards Hahlen, to *b*, *c*.

The outguards in front of Thonhausen, the battalion of grenadiers in that village, and the hussars between Stemmeren and Kuttenhausen, remained in the same position with the detachment at Fredewald, which, however, changed a little the situation of its outguards. The hussar outguard was then moved to V.

The head-quarters at Hille were covered by 2 battalions of British. The Weser being on the left, and on the right a morass, in most places impassable, extending in front of Hille, Sudhemmern, and Hartum, the French army could make no movement unknown to the allies.

§. 65. *Second Instance.*

When Prince Henry, in 1762, was obliged to occupy a defensive position near Freyberg*, the following were the dispositions made for the security of the camp. See Pl. IV.

The position was nearly half way between Nossen and the village of Mulde, with the river Mulda in its front, and Freyberg in its rear.

Rear. Four battalions were encamped behind Freyberg, towards the Spittel forest.

---

* The arrangements for the security of this camp certainly deserve to be noticed here, as they for some time disconcerted all the enemy's attempts, who, however, at length succeeded, in some degree, by their superior numbers, in surrounding the Prussian army.

Two strong redoubts were constructed in the Spittel forest, towards Ober-Schœne and Klein-Schirma, in order to secure the rear, and to be able to resist the enemy for some time in that quarter, with a few troops, in case he should turn the position.

Two regiments of hussars and 4 free battalions under General Belling were stationed at Kleinschirma, with orders to patrol towards Oederan, &c. When the enemy afterwards annoyed the right flank of the army. General Belling placed himself with his detachment farther on, near Langenau, between Gros Waltersdorf and Gros-Hartmansdorf; the hussars were posted on a height and the light infantry farther back, near the forest.

Right Flank. General Kleist was posted at Bertelsdorf, occupying the heights between that village and Weidmansdorf with his own hussars and several free battalions, for the purpose of covering the right flank of the army: he probably had an outguard in front of Weidmansdorf, at III. with a guard of infantry for its support at IV. and during the night, a picquet at V.

The battalion of Luderiz was stationed at Brand. Major Trebra, with his riflemen and 3 squadrons of Kleist's hussars, was at Mœnchfrey: his cavalry outguards are marked at Nos. VIII. and IX. and those of the infantry, on the right and left of the wood, may be traced by the dotted lines. Such, it is imagined, was the situation of these outguards, but their actual position is not known.

When the enemy afterwards attacked the outposts on the right, 2 battalions of grenadiers were encamped between Bertelsdorf and Weidmansdorf, and the cuirassiers of Schlaberndorf behind Bertelsdorf. The communication with Zwickau and Chemnitz, in rear of the right, was at first open.

A regiment of hussars was stationed near the former place, for the purpose of procuring forage in that quarter.

Front. Two free battalions were at Weissenborn, and the Swiss regiment of Heer was stationed near the powder mill, where it was afterwards joined by the battalion of Luderiz from Brand.

A detachment of 100 infantry was placed in the church-yard

at Hilbersdorf; Vorwerk was not occupied in the first instance.

The battalion of Le Noble was at Conradsdorf, and 300 hussars of the regiment of Belling were stationed at Krum-Hennersdorf: the heights from hence beyond Weissenborn, and in front of Weidmansdorf, being occupied by outguards of light cavalry: from the nature of the ground, outguards must have been placed at I. and II, and would necessarily have several detached posts with their vedettes as marked at Conradsdorf, Hilbersdorf and Weissenborn.

Left Flank. Here the regiment of dragoons of Plettenberg was placed *en potence* at some 100 paces distance from the army.

One hundred infantry were stationed at Rothenfurth for the protection of the pass at that place, and that near Gros-Schirma.

An outguard of 100 cavalry was placed in front of Rothenfurth, and 2 battalions were stationed farther to the left towards Nossen, where the communication was still open.

§. 66. *Third Instance.*

The army, when occupying the intrenched camp near Colberg, as shewn in Pl. II. (the trenches *t. o. n. y. z. b. c. c. c. r. h. l.* were formed long afterwards by the Russians) had, besides the guards in the works, one battalion in battery XIX. near the morass, one hundred infantry in the wood at A; a free battalion at Tramp, and several squadrons round Neknin, with a strong guard in the redoubt XXVIII near the morass. The outguards of these posts are marked on the plan.

b) *When the Enemy is several Days' March distant,* the outguards should be posted as pointed out above, but they are not to be supported by the detached parties, picquets, or battalions, which are stationed in villages. They are also weaker and are placed only in the direction towards the enemy, with greater intervals between them.

How far it may be prudent in this case to deviate from the rules, which should be observed when the enemy is near, depends upon the particular circumstances of the moment.

CHAPTER II.

Observations on placing the Outguards.*

§. 67.

The distance between the outguards must be determined by the nature of the ground. They should be distributed in such a manner that the intervals between them may not exceed 2500 paces, when occupying a country which is not perfectly open----that is----small posts should be detached to the right and left from the main guards, so that the average distance may not be more than 2500 paces, and even then, if the whole of the country is to be covered, the furthest vedette will be from 1200 to 1500 paces from the guards, at which distance the report of a pistol, which it is absolutely requi-

* An officer, who is detached, will not require the information contained in this Chapter, with respect to placing the out-guards of a camp, which duty is performed by the general officer of the day; but in posting the outguards of quarters or of a post, it will be indispensible to him.

site should be heard, will, if the wind is contrary, be hardly distinguished.

A guard always risks a great deal, when it is obliged to post its vedettes at a considerable distance, therefore when an outguard occupies a large extent of ground, it should detach small guards.

In Pl. III. the detached post is placed 2000 paces distant from the outguard I. and 3000 from the outguard III. If this post was not placed thus, there would be an unoccupied interval of 5000 paces between the two outguards; and if they should distribute their vedettes so as to cover the whole ground, some would be nearly 3000 paces from the outguard, in which case their firing would not be heard, and during the night the guards would not have timely notice of the approach of the enemy; this might indeed be remedied by directing the whole of the vedettes to fire immediately on hearing the report of a musket, so that the firing would be continued from the extreme vedettes to the guards: but in this case entire dependance is placed upon a single man, and the least accident creates a general alarm.

In Pl. II. the outguards E. F. and G. are posted at much less than 2500 paces from each other, this however the ground renders necessary. If it was adviseable to dispense with the outguard F. the vedettes would be upwards of 1500 paces distant from the guards.

When the picquets are advanced in front during the night, the above rule is subject to many exceptions. In open ground the guards may be more distant from each other, and it is only necessary that they should have a commanding view of the country; and as troops may be distinguished at the distance of 2500 paces, the guards may be posted 5000 paces from each other.

This rule holds good, when near the enemy; when at a great distance, it is not necessary to adhere so strictly to it; but the country should be patroled for some miles in the direction of the enemy, or he should be observed by detached corps----by which precautions you will be in tolerable security.

This exception applies likewise to outguards which are posted by light-troops a considerable distance in front; provided the army is well covered by some other means; as is the case with the outguards I. and II. in Pl. IV. The army is here secured from surprise by the Mulda and the garrisons of the villages, and bridges situated upon that river: the detached post is 4500 paces distant from the guard II.; the vedettes are posted in the day-time so as to have a commanding view of the country, during the night several intermediate ones are added, and those most in front retired a little, so that the whole nearly form a chain.

§. 68.

Of the Distance at which Outposts should be placed from the Camp. No general rule can be laid down for this.

If the outpost of the camp at Hille, Pl. III had been less advanced; as for instance, had they been placed in front of Nordhemmern, the enemy would have been able to annoy and reconnoitre the army in the direction of the villages of Holthausen and Hartum, and the wood adjacent to them; in addition to which, the French army might have approached close to the allies before they could have taken up the position between Hahlen and Holthausen, with their front towards the enemy, as they did on the 1st of August.

The placing of the outposts depends, therefore, upon the particular object in view, and upon local or other circumstances.

If the outposts of the abovementioned camp had been more advanced, as to E. &c. no greater security would have been obtained, for they could not have been supported, and they might have been easily driven in by the enemy.

The outguards F. E. and G. Pl. II. are only advanced about 1000 paces. It is supposed that the ground, at some 100 paces in front of the vedettes, is covered with wood, and that the enemy is very near; besides which, the outguards should, during the day-time, be posted at least 2 or 3000 paces beyond the villages, that timely notice may be given of the enemy's approach. For the above reason, the Saxon outguard (in 1756, when the camp at Pirna was invested) was posted at A. (Pl. II. No. III.) not far from the fortress of Kœningstein. It could not be placed more in advance, as the Prussians were in possession of the wood; from which it was necessarily posted at such a distance, as that the men might be able to mount and form before the enemy could reach it at a gallop; the situation at A. is so near the wood as almost to admit of this, the guard should therefore be constantly mounted, a part only dismounting occasionally to feed their horses at B.

It is always disadvantageous to have the outguards near, unless they are too much exposed by being more advanced. The best situation is the border of a wood, not too distant, in the direction towards the enemy. When the Prussian army, in 1778, was encamped near Nimes, Pl. II. No. II. the hussars stationed at Plauschnitz, had their outguards between Ober and Unter-Wocken, at *b*.

Thus they were above 5000 paces distant from the quarter, for the security of which they were posted: the situation was dangerous, but it possessed a commanding view of the country. Had they been posted close in front of Plauschnitz towards the wood, the enemy might have approached, through the wood, against the troops at Plauschnitz, Neudorf, &c. without danger. Even in the situation abovementioned, the men were obliged to remain constantly mounted; or they would have run great risk of being surprised. A weak detachment in particular should not have its outguard too near, as it is then of the utmost importance to have early notice of the approach of the enemy, and to gain time sufficient to determine upon the measures necessary to be taken according to his force, &c.

When the corps of Platen was posted near Budin in 1773, Pl. I. No. VI. and the Austrian General Sauer commanded a small detachment near Welwarn, the latter placed his outguard at the distance of 6000 paces from him, near Lautzka. This outguard detached a post 3000 paces in advance, and a small party 5000 paces on each flank. As the troops were all light, no danger was to be apprehended from this, and it gave the advantage of discovering the approach of the enemy early.

In such a case, the station of the outguard should be changed, as soon as it becomes dark, in order to avoid being surprised during the night, and that the enemy may not be acquainted with its exact situation.

Therefore, when the enemy is some miles distant, the outguards should be placed at the distance of from 3000 *to* 6000 *paces, if it can be done without danger of their being seized or turned: but when the enemy is near, the outguards must also be nearer, and the troops by which they are posted, whether in camp or quarters, or a detachment, must be held constantly on the alert. If the enemy, being some miles distant, is able to turn the outguards placed at 3 or 4000 paces, and you are not strong enough to post several; a chain af vedettes and small guards may be formed at 1500 paces distant, and at 3 or 4000 paces in front of that, small guards may be posted singly, whose particular duty will be to observe the country from whence the enemy may approach.*

Innumerable instances prove the danger which arises from the outguards being too near. Yet, notwithstanding, the Prussian Regiment of Thadden was surprised. in the campaign of 1778, at Dittersbach, Pl. III. No. II· owing to its having no other guards than those at *a*, *b*, *c*, *d*, *e* and *g*, close round the village. The enemy advanced through the forest A, and the instant he reached the guards, rushed into the village. Had small guards of infantry been advanced 2 or 3000 paces into the forest, and an outguard of cavalry posted as far as *g*, near Neuweisbach, where another Prussian cavalry outguard was placed, this surprise could not have taken place. The Regiment would have had time to turn out and form, after the alarm was given by the firing of the guards in the wood, before the Austrians could have entered the village.

Several instances of this kind are to be found in the Section treating of surprises. §. 57, &c.

§. 69.

The Situation of the Outguard should be such, that it may, at a distance, be concealed as much as possible, from the enemy.

The best situations are in pits, behind woods, heights, hedges, &c. provided the spot is not surrounded by this cover, and that the movements of the guard are free from impediment.

The outguard at Thonhausen would have been well placed, during the day, behind the wood, on its right; but, during the night, outguards should be placed in open spots, where they can discover every thing round them; and they should always at night be posted in a different place from that they have occupied during the day; so that if the enemy attempts a surprise, he will find a vacant spot, and the guard will have time to take such measures as may be necessary.

§. 70.

The Strength of the Outguard depends, first, on the number of vedettes it is required to furnish, and secondly, on the resistance it is intended to make.

Each outguard should consist of three times, or at least double, the number of vedettes and sentries, which it is to post. Therefore, if 10 men are required for the vedettes, the outguard should consist of at least from 20 to 30.

Vedettes, which are posted at a short distance, consist only of one man; those at above 1500 paces, when the strength of the guard will allow it, of 2.

The strength requisite for resistance, cannot be determinately fixed. If the enemy is near, and purposes to reconnoitre the camp, scarcely any outguard will be strong enough to hold him in check.

In such cases, the outguard may be supported by a battalion, or a detachment of infantry, if the ground is intersected.

The outguard in front of Thonhausen could have fallen back, in case of necessity, upon the battalion stationed at that place.

The cavalry picquet, *b*, between Hahlen and Hartum, was supported by the infantry picquets, at Hartum; and by that stationed at *b*, at the cross road near that village.

Outguards are generally reinforced, when near the enemy, both for the purpose of posting of more vedettes, and of opposing more effectual resistance.

CHAPTER III.

General Rules for the Conduct of Outguards of Cavalry and Infantry.

§. 71.

Previous to the Guard or Detachment marching off, the rules prescribed in §. 25, should be observed.

§. 72.

Division. As soon as the outguard has arrived at its post, the officer will proceed to dispose the sentries and vedettes, taking care that the most vigilant and trustworthy men are placed at those points where danger is most to be apprehended; and that in the double sentries and vedettes, an old attentive soldier is always joined with one less to be depended upon. The guard should be divided into 2 or 4 divisions, according to its strength, each to be commanded by a non-commissioned officer. If it is one of the common outguards of the camp, its situation, together with that of the sentries and vedettes, will be determined, in the first instance, by the general of the day. But the officer commanding is at liberty to increase the number of his sentries at night, or make such alterations as he may conceive necessary for his security, according to circumstances, reporting the same to the general of the day.

§. 73.

Knowledge of the Country. If the post has not been occupied before, the commanding officer should send for some person, from the nearest village, who is acquainted with the neighbouring country, and ride with him to the distance of 2 or 3000 paces round, in order to gain a knowledge of it himself. He should mark down the names of the places, rivers, &c. and inquire the distance of the surrounding villages and woods; he should observe the position of the stars; and thus render himself competent to give information to the generals who are reconnoitring; and also if any thing takes place to particularize the spot where it occurred. He should judge the distance of the enemy by the distance of objects, &c.

If the post has already been occupied, he will require the information above stated, from the officer whom he relieves, and with whom he will inspect the ground on horseback. If the enemy is supposed to be concealed in the country, the reconnoissance should be made before the old guard marches off, as there is then double the strength on the spot.

§. 74.

Instructions to the Vedettes and Sentries. The commanding officer delivers the orders to the vedettes and sentries himself; but he should not be contented with that, and should examine them, to see if they are acquainted with the duty they have to perform.

Peasants or Travellers arriving singly, if they wish to pass the country occupied by the posts and guards, should be directed, by the sentries, to the guard, where they will be examined by a non-commissioned officer.

When covered Waggons, People in a Body, or armed, an Enemy's Officer, or Trumpeter, several Deserters, or a Party of the Enemy, are perceived approaching; one man from the double vedettes, reports the circumstance to the guard, while the other remains watching them, and orders them to halt when they arrive about 100 paces from his post, and wait until they are examined by a party sent from the guard. If they do not obey, when ordered to halt, the vedette must immediately fire.

When any Movement is observed during the Night, the vedettes or sentinels demand *Who's there;* if the answer is: *Patrol*, they require the countersign; if the answer is not *Patrol*, the vedette calls *Halt*, and immediately gives notice to the guard that some person is approaching. If there is reason to suspect, from his speech, dress, &c. of the person, that it is an enemy; or, if upon being ordered he does not halt, the vedette will fire. See §. 87.

The Space, to which the particular Attention of the Vedette or Sentry is to be directed, should be pointed out to him. He also observes the vedettes on either side of him, and gives immediate notice to the guard, if he misses one. If a vedette perceives the enemy in his rear, he fires immediately.

On the Advance of the Enemy, the vedettes and sentries fall back, the moment the alarm shot is fired: if the country is open, they retire upon the guard, but inclining towards the flanks, so as to leave its front open.

§. 75. *Conduct of the examining Party and of the Guards, with respect to Persons approaching.*

Every post or guard has a subaltern and a party a short distance in advance, who are constantly on horseback or under arms; or has at least a double vedette. A party, appointed for that purpose, or a non-commissioned officer, with a few men, examine carefully every one who arrives.

Peasants or Travellers, who have business in the neighbourhood, may be allowed to pass through the camp, if there is no particular order to the contrary.

Pedlars, Horse-Dealers, and Persons with Petitions, were not allowed to enter the camp of the allies, during the seven years' war.

If any thing suspicious is observed about a traveller he must be detained, and his detention reported to the general of the day.

If an Officer or Trumpeter from the Enemy, or a Party of Prisoners arrive. The officer of the guard, after their having been examined by a non-commissioned officer, repairs himself to the spot, where they have been halted, and receives the letters or prisoners, together with an authentic return of them, and gives a receipt for the same. The officer or trumpeter is conducted a certain distance back by a small escort, and what has been received is forwarded to head-quarters.

If an officer wishes to proceed to head-quarters, he must remain at the first station until he receives permission; which being obtained, he is blindfolded, and conducted thither by a serjeant and 2 men.

If Deserters arrive successively to a considerable number, they should, in the day time, be forwarded to the camp, at night they should be guarded aside, or if the guard is not numerous, be sent back. Their arms should be taken from them before they are admitted within the chain of posts, and no person should be allowed to converse with them. The officer commanding the guard should make a memorandum of their names, the corps to which they belonged, the place whence they came, the enemy's strength and situation, whether he is sickly, and how provisioned.

When a Detachment arrives, it is, in the first instance, examined by a non-commissioned officer with a few men, who demands the field-cry, this, however, can only be given, if the detachment has marched out during the same day; in either case, the commandant of the detachment is conveyed to the commanding officer of the guard, who requires from him a pass or other written document. If nothing suspicious is discovered, the party, which has hitherto remained where it was at first halted, is permitted to proceed; the guard being mounted, or if infantry, under arms. If the party consists only of a few private men, it should be ordered to remain until day-light upon one of the flanks.

§. 76.

Reports. All certain information respecting the enemy, which is received at any outpost either of infantry or cavalry, is to be reported immediately and at the same time, both to the field officer of the picquets and to head-quarters, to the latter the report is to be made in writing. It should be stated whether any thing has been observed from the post itself; the place, hour, name, and rank, should also be specified.

During the seven years' war cards were made use of for reports.

When posted near the enemy, particular attention must be given to observe and report his movements; this is the principal object of an out-post. A written report is sent by a non-commissioned officer to the field officer of the picquets, an hour before the retreat.

§. 77.

The *Parole, Field Cry, and Countersign*, are transmitted, sealed up, by an orderly man from the field officer of the picquets to the officer commanding the outpost. The field cry is communicated to the non-commissioned officers, and the countersign to the privates. If a post is detached quite out of communication with the army, the commandant gives out the field cry and countersign, and if there are several officers with it, the parole also.

CHAPTER IV.

Disposition and Duty of Outposts of Cavalry—Placing of Vedettes.

§. 78.

Double Vedettes and Intermediate Vedettes. All Cavalry outposts should have double vedettes during the night: and between the extreme vedette of the chain and the guard an intermediate one, (which should consist of 2 men if the guard is strong enough) should be posted obliquely, so that if the enemy attacks one, the other may fire upon his flank; and if the former is carried off, the alarm will be communicated by the latter.

§. 79.

Distance of the Vedettes, and Patrols of Vedettes. The distance of the vedettes from the main guard and from each other should be such that they are not in danger of being cut off, and that they may obtain as extensive a view as possible of the country round. It is seldom right to post vedettes more than 1500 paces from the outguard; if they are at a greater distance, their firing sometimes cannot be heard. The manner of posting vedettes is more particularly described in the subsequent examples in Chapter V.

If the vedettes are placed so far from each other, during the night, that the enemy may easily penetrate between them unobserved; one of the men, when the vedette is double, should patrol towards the next. In this case the vedettes should be relieved every hour; and the intermediate spaces between them should be occasionally observed, to ascertain whether the patroling is regularly performed.

§. 80.

Stations of Vedettes—relieving them. Consideration of the Ground. Single Vedettes posted in the Rear. Vedettes should be relieved frequently during the night, unless it is attended with much danger. The vedettes are posted, in the day-time, on the hills, and at night, in the valleys and passes. If a vedette is posted in front of a defile, he will, at night, be drawn back either into or behind it. A vedette, who is stationed on a height which overlooks the neighbouring country, at from 1500 to 2000 paces in advance, should be retired during the night and placed in a line with the other vedettes. At night single vedettes should be posted at a short distance only on the flanks or in the rear of the guard, in order that they may be secure in case of danger.

§. 81.

Duty of the Vedettes at Night. This subject has been already treated upon separately, under the head of General Conduct----where the duty by day is also pointed out.

The manner in which vedettes should conduct themselves, on perceiving any body approaching, cannot be too frequently impressed upon them. They immediately demand aloud, (so as to be heard all round) "*Who goes there?*" and one man advances a few paces, with his piece cocked, and orders whoever is approaching to *Halt* and give the counte sign. (When the patrol answers to the challenge they give the counter-sign without it being demanded.) If the countersign is properly given, still the party approaching must halt, until the man who advances has rode up and convinced himself that it is not an enemy. When the vedette discovers that it is an enemy advancing, either from the countersign not being given, or from his speech betraying him, or if he moves after being ordered to halt, or from another party approaching on the opposite side; the foremost man fires and retreats behind the second, who in the mean time fires also. The whole chain of vedettes fire at the same time. If the party approaching the vedettes does not give the countersign, but halts and remains at a distance, this should be reported by the second man of the vedette to the guard, from which, as it is pointed out in another part, a non-commissioned officer should be sent with two men to ascertain the cause of this circumstance. At night the vedettes should have their pistols in their hand, and in the day, their carbines resting across their saddles.

§. 82. *Duty of the Guard at Night.*

Orders are often given that half the outguard should be on horseback during the night, and are as often neglected. The rule in this respect is, that the party which is mounted should be sufficiently advanced to give time to the others to mount also. A spot should be previously pointed out for the dismounted party to form, in case of the outguard being attacked, so that when the other party has fired and is retreating, they may throw in their fire with good effect. If in cold bad weather, the outguard has a fire* at night, it should be in some covered spot in the rear, and the men should only go there in turn, a few at a time, to warm themselves.

§. 83. *Conduct when an Outpost is surprised. Conduct when pursuing or when retreating.*

Notwithstanding all the above-mentioned precautions, an outguard may be surprised and unable to retreat in good order. In such a case the men should take the bold resolution of throwing themselves impetuously and sword in hand upon the enemy, shouting loudly at the same time; it is probable he will fly, but as he may be strong, he must not be pursued. It will be necessary, however, to fall back immediately and take up a more secure position. The best disposition, when a retreat is unavoidable during the night, is to form the outguard in 2 or 3 divisions, which should retire at a short distance one after the other. The men being formed in single rank.

§. 84. *Disposition of the Guard to prevent Surprise.*

The outguard should, during the night, establish

* Many old and experienced officers disapprove of a fire, conceiving that it renders the men not only sleepy, but unhealthy, and that they can in general, secure themselves better from cold by exercise.

two small posts about 300 paces to the right and left. This will render it extremely difficult to surprise the main post, and if the enemy should advance to attack it in front, these small parties will be able to check him by their fire upon his flanks. The Prussian out-posts before Troppau, in the winter of 1778-9, adopted this disposition, in consequence of having been frequently attacked by the Austrians.

§. 85.

Duties of the Detached Parties and Vedettes, on being attacked by the Enemy. A non-commissioned officer's party detached in front of an outguard should retreat on one side, and not direct upon the main guard, for if they do so, instead of gaining the enemy's flank, the whole will be thrown into confusion. They should, however, only fire, and retire immediately the enemy advances. It must be strongly impressed on the vedettes that they should also in the same manner retreat upon the guard, for the reasons already stated.

§. 86.

Duty of an Outguard in the Day Time. The horses must not all be unbridled at the same time, during the day. If the guard consists of 20 men, one half of the horses may be kept unbridled, and one third when the guard is stronger. The non-commissioned officer's party, which is advanced some hundred paces in front, for the purpose of reconnoitring, should remain constantly mounted.

§. 87.

Conduct on the Approach of the Enemy during the Day. Retreat by Day. If the enemy advances during the day, the officer commanding the outguard must (as far as possible) ascertain his force, and if he is weak, will place his men in ambush and endeavour to cut him off; but, on the other hand, if he is strong, which it is most probable he will be, a small party must be sent to the rear as a reserve, and the remainder of the guard must be formed in 3 divisions. If, in this case, the guard is hard pressed in retiring, the reserve should shew itself immediately, which the enemy will probably conceive to be stronger than it really is, and will retreat. If the guard retreats in 3 divisions, without having any reserve, the flank divisions should remain at a short distance from the centre, and cover its retreat, if it is obliged to give way; or they may conceal themselves, in order to appear unexpectedly. If the enemy pushes on rapidly before the picquet can turn out and form regularly, he must be instantly attacked, and, if possible, on the flank.

§. 88.

Of relieving the Outguards. When near the enemy, the outguard should be relieved an hour or two before day-break, until which time, and till the vicinity of the post has been reconnoitred, the old guard should remain on the ground.

CHAPTER V.

Duty of the Patrols sent out by Outguards.

§. 89. *Explanation.*

Patrols are sent to a greater or less distance from the outguard.

a) The latter patrol within and without the chain of vedettes, to see that the men are attentive, and that the enemy has not approached too near, or penetrated between them.

b) The former proceed to those points whence the enemy may advance, and every morning carefully explore the intersected ground, to prevent any party of the enemy, that may have concealed itself during the night, from making an unexpected attack on the guard, or reconnoitring the country.

§. 90. *Patrols to a short Distance.*

Strength and Time. They are from 2 to 4 men strong, and are employed principally during the night.

Disposition. a) When a patrol visits the vedettes, the non-commissioned officer is in front, and 2 men follow at the distance of about 30 paces, that they may be ready to fire, if the non-commissioned officer suddenly falls in with the enemy between the vedettes.

b) When the party patrols in front of the vedettes, one man is sent in advance, and is followed at about 100 paces, by the non-commissioned officer who commands the patrol and 2 men, the fourth being at an equal distance further in the rear.

This distance should be increased or diminished according to the darkness of the night. The leading man must look frequently round, that he may not be separated from the non-commissioned officer and men in his rear, and who should examine the bushes, &c. to the right and left.

Road. These patrols proceed in intersected grounds from 100 to 200 paces, and in an open country, about 600 paces in front of the chain of vedettes. The officer commanding the guard points out in the day time to the non-commissioned officer, the exact direction which the patrol is to take.

Both kinds of patrol proceed sufficiently far to fall in with the vedettes of the next post. Thus the whole ground between the two outguards is well examined, and the vedettes by this means kept on the alert.

General Regulations. It is usual to patrol with the vedettes that have been relieved; they are formed in 2 divisions, and immediately after the relief a serjeant performs the patrol above-mentioned in front of the vedettes, with the first division; this is necessary to discover any enemy that may have been concealed for the purpose of observing the stations when the vedettes were relieving. About half an hour afterwards a patrol is made, with the second division, within the chain of vedettes. Thus the vedettes being relieved hourly during the night, the front of the chain will be reconnoitred thrice during each hour; once by the relief, and twice by the patrols.

When two Patrols meet, they demand from each other the *Field-Cry*. On the challenge, *Who goes there?* the first man advances with his pistol cocked, and on receiving the answer *Patrol*, orders it to *halt*. The non-commissioned officer who commands the patrol, then advances towards the party that has halted, and demands the *Field-Cry*, which is given to him by the commander of the other patrol, who has advanced likewise.

When a Patrol falls in with the Enemy, it immediately fires, that the guard may be instantly alarmed.

On arriving at the Guard, the patrol is halted and reconnoitred by the vedette stationed at that point, and by a non-commissioned officer and 2 men. The vedette calls out *Who's there?* and, on *Patrol* being answered, *Halt*. The non-commissioned officer who is to examine it, demands the *Field-Cry*, and at the same time advances with his piece cocked, and receives it from the leader of the patrol.

§. 91. *Patrols to a greater Distance.*

Strength. When patrols of this description advance, during the day, through an intersected country to a considerable distance, and are obliged to leave one or more defiles in their rear, they should not consist of less than 12 men; otherwise 4 or 6 men will be sufficient. If they have to patrol large woods they should consist of 20 cavalry or more.

Time. These patrols should be sent out more particularly towards day-break, as the enemy can then only attack in a body, unless he incurs the risk of being thrown into disorder upon ground with which he is unacquainted.

A strong patrol is therefore sent out at day-break to examine the adjacent woods and villages.

Direction and Distance of the Patrol. a) The patrol proceeds towards the quarter, whence an attack is to be apprehended.

b) It advances, if possible, through woods, where information respecting the enemy may be collected, without allowing its object to be known, or affording an opportunity to the enemy to cut it off.

c) It proceeds as far as it can without incurring the danger of being cut off. Frequently from 4 to 10 miles.

The Disposition and Duty of these Patrols have already been pointed out in the section relative to patrols. §. 1, &c.

CHAPTER VI.

Examples of the Conduct of Outguards and Detachments in the Chain of Posts.

§. 92. *Cavalry Outguard in Front of Thonhausen.*

(Pl. III No. I.)

The main guard at I. is dismounted, and the horses are fed by turns, a little to the rear, near Thonhausen; one double or two single vedettes are stationed on the road towards Minden, near the Weser; in the rear of which a corporal and a private stop, at a certain distance, and examine all persons who approach along the roads, or who appear farther in front, or to the right.

At night, the two men composing the vedette are posted, one on each of the roads, and a third is placed near them on the Weser. One man is also stationed, during the night, on the Weser, near the outguard; lest the enemy should be able to cross any ford, which is not known to the outguard, and surprise it. Lastly, a single vedette is placed in front of the small wood to the right, about the spot where the source of the little stream is marked.

To the right of this guard is a detached post, consisting of at least 12 men, which has two double vedettes, as shewn on the plan. At night these two double vedettes are formed into four single ones, who are placed at the distance of 500 paces from each other. When the post is sufficiently strong, double instead of single vedettes should be employed in this as well as in the above cases.

The patrol, which is sent out at day-break, consists of from 12 to 24 men, if the strength of the guard will allow it; two men proceed to the right, and 2 to the left, (by the houses and gardens, which are in front of the vedettes of the outguards) and 8 pass along the road between them. The 2 men on the right search the other houses and gardens on that side, towards Hahlen. The 8 men in the centre proceed to E on the heights, and the 2 on the left observe the road towards Minden.

Three men afterwards reconnoitre the country as far as E every hour.

The patrol of 3 men proceeds every half hour, during the night, as far as the first gardens, and there turns to the left down to the Weser. The detached post sends its own patrols between H and V, which however do not pass through the narrow passages between the gardens.

§. 93.

The hussar outguard, Pl. III. in front of Holthausen, has, as long as the enemy retains possession of Hille and Hahlen, 3 double vedettes in the day time; one of which is posted near the wood, either on the edge of the open part of it, or, which is preferable, 200 paces from it At night, 2 more double vedettes are placed between the 3 already mentioned, or the double vedettes will be formed into 4 single ones, which are at equal distances from each other. The distance between the vedettes will not then exceed 500 paces.

The patrols from this outguard examine the small woods between the guard and the villages of Hartum and Hahlen every morning, and even enter these villages, if the enemy will permit it. A patrol of from 3 to 6 men proceeds, in the first instance, towards Nordhemmern, and from thence as far as Sudhemmern; about 8 or 10 minutes afterwards, another patrol is sent in 2 parties into the small wood; the one party moving to the right, and the other to the left of the road, near which the centre double vedette is marked in the plan; after searching the wood, one party turns towards Hahlen, and the other towards Hartum; a few men only enter the villages, the remainder halt about 300 paces in front. This is the common morning patrol; its strength cannot be fixed, but should be as great as the numbers of the guard will admit. After this, one patrol of 3 men only proceeds through the small woods, and then reconnoitres the villages.

The villages, with the exception of Nordhemmern, are not reconnoitred during the night; the patrols passing only through the small woods and across the heath, at the distance of about 500 paces in front of the vedettes. In Nordhemmern the patrol proceeds no further than the church, at least, only 2 men advance beyond, while the third remains in front of it.

Such was the disposition, while the enemy still occupied Hille, and occasionally shewed himself at Hahlen. But after the French hussars had been surprised at Hille on the 20th of July, and that place and also Hahlen were evacuated, it became practicable to make a new disposition for this outguard at III. and to place vedettes at *r*, *s*, *t*, and *u*, on the border of the wood, in such a manner that they might command an extensive view of the country, without being perceived by the enemy. But in order that the vedettes may not be too great a distance from the guard, it is necessary that it should be moved forward into the wood as far as the cross way marked *b* on the plan.

During the night the guard and vedettes resume their former positions, or a small post is detached in front of the wood, on the road between Nord and Sudhemmern, which posts a double vedette at Nordhemmern, and other single ones intermediately between those already posted. It is however safer to resume the former stations during the night.

§. 94. *Hussar Outguard near Hilbersdorf.*

(Pl. IV.)

The station of a hussar outguard of 45 men, near Hilbersdorf, is shewn in Pl. IV. posted for the purpose of guarding the country between Hilbersdorf and Lichtenberg, and consequently to 4 miles in front of the right wing of the Prussian army.

This cannot be effected by a single guard, as the vedettes would be at too great a distance; and as this distance should not exceed 1500 or 2000 paces, even 2 guards would not be sufficient to cover an extent of 4 miles or 9000 paces. The main guard, therefore, should be placed at II. in the centre of the ground, having a detached post on each flank.

The main guard has 1 single and 2 double vedettes, as shewn in the plan. The double vedette near *x* is above 2200 paces from the guard, but may always be seen from thence, and if nearer it would not have a sufficient view of the country; 2 trusty men should always be selected for this service. The detached post between Weisseuborn and Lichtenberg, has 3 double vedettes, 2 of which are at such a distance from the guard, that a shot fired by either of them could not be heard there, nor can they be seen from thence: for this reason, an additional vedette is placed at *y*, who must keep his attention constantly fixed on the former, and fire when either of them fires, &c. If these vedettes were not placed at such a distance, they would not obtain a sufficiently extensive view of the country.

The second detached post is stationed in front of Hilbersdorf, and has only one double vedette, which is placed so as to have a good view of the low ground.

The post between Weissenborn and Lichtenberg must, at night, draw back its vedette to *y*, on the front of the hill, where the *poste d'avertissement* stands, and place it 200 paces on this side of the trees. The single vedette, which was at *y*, must be removed to *z*, and the double one drawn back towards Lichtenberg as far as *t*. But as the vedettes are still farther distant from each other than is prudent at night, 4 single should be formed from the 2 double vedettes. The extent *z*, *y*, *t*, as far as the Mulda, will then be covered by 4 vedettes, at not more than 600 paces distant from each other.

The best situation for the guard at night will be near V, in order that any party of the enemy, purposing a surprise, may not immediately discover it. If then the ground is patroled 600 paces in front of the vedettes, the country round the village will be guarded to the distance of 1500 paces. By this disposition, the ground is well covered, and the outguard secured from surprises. But it will be requisite that the two batalions at Weissenborn should be drawn into the church during the night, as otherwise the enemy, approaching by way of Lichtenberg, might enter the village before they could turn out. For the security of these battalions, the guard would be best placed at *z*, and its vedettes in the line *w*, *w*, *w*: but in that case, it would be necessary to place an infantry guard at Lichtenberg, at *s* and *s*, during the night.

The main guard at II. remains in the same station during the night, and places an intermediate vedette between it and those that are distant, but forms them single, that it may not be in want of men. The vedettes should be pushed forward as far as the foot of the hill during the night, and posted on the dotted line *w*. *r*.

The patrols are sent about 400 or 500 paces in front of the line *w*, *w*, *w*, *r*. And near the 2 guards, where the vedettes are single and distant from each other, a patrol along the line of vedettes should be made every half hour.

The morning and day patrols must search the forest in front;

and for this purpose, the morning patrol should consist of at least 20 men, that it may be able, without danger, to reconnoitre the whole, at the same time in several parties; and this should be the case, even if a non-commissioned officer and a few men only remain to observe the vedettes.

It is unnecessary to say any thing respecting the guard detached towards Hilbersdorf, the conduct to be pursued by which, may easily be learnt from what has been said above.

§. 95. *Cavalry and Infantry Outguard near Weidmansdorf.*

Near Weidmansdorf 100 infantry and 24 cavalry are posted, for the security of a corps (detached from the army near Freyberg) stationed near Bertelsdorf.

As long as Mœnchfrey remains unoccupied by the hussars and yagers, who were afterwards stationed there, the following disposition may be made for the above-mentioned detachment:

During the day, the cavalry outguard is placed at III. Pl. IV, and has 2 vedettes, who possess from thence an extensive view of the country. The infantry are stationed in the village of Weidmansdorf, near IV, and post 3 double sentries.

The cavalry post will, at night, be drawn back to VI, and a chain of 7 single vedettes will be formed on the line *zz*. In the wood near *v*, a post of infantry from Brand is placed.

The post of infantry at Weidmansdorf remains at IV, and detaches a small post to *y*, and another to the left of the Mulda; it also places double or single vedettes, as may be necessary, from *z* to that river, at 200 ot 400 paces from each other, according to the darkness of the night.

When the enemy is not very distant, the patrol should first search the whole neighbourhood in the morning. The cavalry outguard then makes a patrol towards Helbigsdorf; while a party of 12 infantry reconnoitre the banks of the Mulda, as far as Hilbersdorf. As soon as the cavalry guard has passed Mudisdorf, it detaches one party to III, on the left of the hill, and another to the right, as far as the hill marked VIII; and when it has arrived at this point, the main party proceeds beyond the line *xx*, as far as the village, which must be examined. Two patrols of 3 men each will then be sent to reconnoitre the wooded hills beyond the village, and the outguard resumes his station at III.

During the night, when this guard is stationed at VI, it sends the above-mentioned patrols of 3 men each, as far as Mœnchfrey, from whence they proceed to VIII, VII, and III, but never enter either the wood near Mœnchfrey, or that on the hill, as they might be seized by the enemy, and could not, at all events, accomplish their intention of searching them.

When a corps of yagers and hussars is posted at Mœnchfrey, as was afterwards the case, in 1762, the outguard must not retire farther from III, than from the hill, nearer to the village of Weidmansdorf, and the vedettes must then be placed in *x*, *x*, *x*, *x*, *x*. The infantry guard will act as above, namely, will divide itself into 3 posts.

§. 96. *Corps of Cavalry and Infantry near Mœnchfrey.*

This corps was stationed at Mœnchfrey, in 1762, when the enemy was very near, and appeared anxious to undertake some enterprise in that quarter.

This detachment had consequently reason to apprehend a surprise every night. A chain of posts at the distance of 2000 paces might in some measure afford security against a surprise by night; for although the patrols should not discover the enemy before he arrived close to the posts, yet the detachment would be able to get under arms, after hearing the first shots fired, before the enemy could advance 2000 paces, opposed by the picquets and delayed by the darkness. A cavalry guard is posted to the left near Mœnchfrey, and another near Langenau. There is also an infantry guard to the right and another to the left, in the wood, as shewn in the plan.

The cavalry guard, VIII, moves forward at night to the spot where its vedette stands, and places its vedettes as far as Helbigsdorf, 400 paces in its front. The cavalry guard, IX, keeps its station during the night, and draws its vedettes behind the village.

Notwithstanding, that by this arrangement, the enemy will be discovered at about 3000 paces from Mœnchfrey; yet it will be adviseable at night to post, at from 2000 to 3000 paces, in front of the outguards, single parties of cavalry, of from 3 to 9 men each, in the open ground; and of infantry of from 4 to 12 men each, where it is intersected, at the places by which the enemy may approach.

These parties will keep off the enemy's small patrols, which might creep within the posts, and will discover his intentions without his being aware of it, which indeed must be their principal aim. The situations in which they are to be placed, should be fixed during the day, but not occupied until it is dark, that the enemy may not be acquainted with them.

They should conceal themselves in covered spots, 2 men being always together, at intervals of 600 paces, so as to form a line, through which it may be difficult for the enemy to pass undiscovered. In bad weather, they should be provided with cloaks, blankets, or some other sufficient covering. On perceiving the enemy, they should fall back without firing, in order to give intimation of his advance to the guards and to the corps. If one man is suddenly attacked and obliged to fire in his defence, the remainder must fire also, but as this may not be heard by the guard, some must convey thither immediate intelligence.

§. 97. *Detachment of Cavalry and Infantry at Fredewald.*

(Pl. III.)

This detachment consists of 300 infantry and 100 cavalry: 100 infantry are always on guard during the day, and the remainder are quartered in the houses. The 2 outguards marked on Pl. III. consist of 30 men each; the other 40 are stationed in the village, in the house where the commanding officer is quartered, and which is the general rendezvous. They have a

Poste d'avertissement upon the tower, and both by day and night, several sentries at the distance of 300 paces round the house, in which is the commanding officer, and during the night, the remainder of the detachment. Of the cavalry 32 men are on guard every day, 16 of whom are constantly at the commanding officer's quarters, with their horses bridled; an outguard is formed with the other 16, and posted on the hill in front of Dutzkamp. At night, a picquet of 50 infantry reinforces the outguard, and furnishes a chain of sentries round the village, at the distance of 1000 paces.

The cavalry outguard will be drawn back as far as VI, and will post 1 single and 2 double vedettes; the remaining 16 will be posted as an outguard at IV, and replaced at the commanding officer's quarters by a picquet of the same number, who will have their horses constantly bridled. Thus half the detachment is always on duty at night. The remainder are quartered in the houses nearest to that of the commandant. A light must be kept, and one man remain on the watch during the night in each house, in order that they may turn out more readily in case of alarm.

§. 98. *Posts of Infantry in Defiles or Woods.*

In Pl. II. No. III. a post of infantry is shewn at *a*, belonging to a Prussian corps, which is more to the right, and which, in 1756, was stationed near the Elbe, on the investment of the Saxon camp, near Pirna. A Saxon outguard is posted at A, in front of Kœnigstein, the farthermost point of the Saxon camp. The post of infantry at *a*, may be turned through the wood; in order to prevent this, several sentries must be posted towards *b*, at 600 paces from each other by day, and 300 at night, until they form a chain with those from the redoubt *b*. The guard will block up the road along the Elbe with some felled trees, and will place itself in the small wood; but as the impediment on the road cannot be easily removed at night, it will then be necessary for a small part of the guard only to remain near it, while the remainder move rather forwarder into the wood, in an oblique direction. This is more necessary if the guard is not sufficiently strong to post sentries at 300 paces from each other, as far as *b*. If one part inclines, as above mentioned, to one side, the enemy having turned the guard, will not be able to discover it, and it may, perhaps, be able to fall upon his flank unexpectedly.

A common road may be rendered impracticable, by digging a ditch, and laying some trees, not too large, across it. But the surest method is to raise a pile of wood in your rear, and set fire to it, when it becomes necessary.

§. 99. *A Guard of Infantry in a Wood.*

In Pl. IV, a post of infantry is placed in front of Mœnchfrey, in a wood to the left, which extends close to the enemy. The guard stands in the centre of the wood, and has 3 vedettes at not more than 1000 paces distant, and about 500 paces from each other. These vedettes would have a sufficient view of the enemy, should he advance in the day; but at night they are too distant from each other, and an additional vedette should therefore be placed in each interval.

The guard, for its own security, removes at night, sometimes to the right and sometimes to the left; so that the enemy, with the best information, cannot be certain where to find it.

A guard of this description should always send out a few patrols of 2 or 3 men each, who should advance silently about 1000 paces in front of the sentries; they should listen frequently and attentively, and should remain some time in the same place. Half of the men must always be on foot, with their arms close at hand, and the remainder laying down beside them.

In Pl. II. No. III, a post of Prussian yagers is placed at *d*, and although consisting only of 20 men, has to guard a space of 3000 paces. In this case, the only thing to be done, is to form a path of communication through the wood, and thus to regulate the distance and the men: 6 men will be allotted to each 1000 paces; if 3 then are always on sentry, the distance between them will be 500 paces. But as the enemy might, notwithstanding, pass undiscovered through the intervals, a constant patrol must be made with 2 men.

This disposition is by no means disadvantageous, when the guard is weak, and the space extensive; as it can only be considered a *Poste d'avertissement*, and it is impossible for the enemy to carry off the whole of it.

The duty of guards in entrenchments, or defending a ford or bridge, will be detailed in the part treating of field fortification.

SECTION V.

Of the Conduct of Stationary Detachments.

CHAPTER I.

Conduct of a Detachment of from 10 to 30 Men, which is to remain a certain Time in a Country, for the purpose of observing it.

I. CAVALRY.

§. 100. *Explanation.*

When an army is encamped, or in winter cantonments, small detachments are usually posted in its front towards the enemy, at the distance of from 2 to 8 miles; which have no other parties upon their flanks, at least within the distance of 2 or 4 miles, and must therefore depend for their security upon their own vigilance and circumspection. The object of these detachments is not to defend, but to watch the country, and give immediate intimation of the advance of any corps of the enemy.

This is particularly done when an army is in cantonments. A chain of posts, of hussars, or dragoons, is then formed at 2 or 4 miles distant from the quarters nearest to the enemy, and at 1 or 2 miles from each other. They are generally relieved every 3 days, and are not confined to any particular spot, but are to observe a certain space of country, which is pointed out to them. Corps are stationed for their support at 2 or 3000 paces, or sometimes 4 miles in their rear.

In this case, the outguards or detachments should be posted more in advance than when an army is encamped, as more time is required for the troops to assemble.

§. 101. *Supposition.*

Suppose 30 cavalry are detached about 6 miles distance from the army, to reconnoitre a country partly open and partly intersected.

§. 102. *Observations previous to marching.*

The detachment should be supplied with as much provision and forage as possible. You should make yourself acquainted, by every enquiry, with the road by which you are to move, and the distance of the enemy. In other respects pay attention to §. 25.

§. 103. *Observations on the March.*

Even if you have intelligence that the enemy is several miles distant, still as you are uncertain whether he is not in the country through which you must pass; every precaution should be observed during the march.

a) An advanced guard must be formed, which should be, in the day time from 1000 to 2000, and at night, 2 or 300 paces in front of the detachment; and, as the object is to observe the country, it is advantageous to throw out flankers, who will move by the roads to the right and left, and make enquiries respecting the enemy.

If the detachment consists of 30 men, as is supposed in this instance, 6 should compose the advanced guard, and two parties of 3 men each should advance upon the by-roads to the place of destination.

b) The advanced guard and flank parties observe the same conduct as the patrols. They must not pass any village or defile until it has been searched by a few flankers. The adjacent woods and bushes cannot indeed be examined, and it is therefore necessary, with respect to them, to rest satisfied with the information of the country people, that the enemy has not been seen in the neighbourhood.

c) But if the information received is to the contrary, every precaution must be used both by the flank parties and the advanced guard. The woods between the army and the posts must be carefully reconnoitred; and the detachment must not advance through a defile, (by which it is absolutely nesessary that it should return) before every spot within 2 or 3000 paces, both in front and on the flanks, in which the enemy might be concealed, have been thoroughly examined by a few men.

d) Under all circumstances, the commanding officer is to make himself acquainted with the country, the defiles towards the army, and the impracticable parts, such as rivers, morasses, and steep mountains. And he will find it extremely advantageous to be furnished with a map, which he may correct, where necessary. He must also mark down the names of the places, woods, rivers, rivulets, &c. that he may be able to make a faithful report, when called upon.

e) He will take guides from the villages, and will procure from them information respecting the country; particularly the fords on the rivers, the roads across the morasses, and the passages of the mountains.

§. 104. *Arrangements to be made for the Security of the Detachment on arriving at the Place of its Destination.*

a) *Outguard.* If the detachment arrives during the day, an outguard must be instantly posted, which remains mounted, while the detachment feeds its horses in some concealed spot.

b) *Reconnoissance.* After the horses are fed, the commanding officer proceeds to reconnoitre the country; for this purpose, the detachment advances in 2 or 3 parties, to the distance of 2 or 3 miles. The commanding officer previously points out to the non-commissioned officers, the places to which they are to patrol. He shews them the roads by which they are to advance, and explains to them the precautions to be observed respecting the ground in front.

c) *Conduct in the Day Time.* The above duty having been performed, the detachment will retire to some convenient spot, and there *bivouacque* during the day; unless the weather is either extremely hot or rough, when a barn, well situated for this purpose, should be sought out. If the cold should render a fire absolutely necessary, the detachment must enter some village previously fixed upon. Here an alarm post must be established, where the whole must rally, the moment a shot is fired; it should be towards the rear of the village. This alarm post also should be pointed out to the outguard, who must retire upon it by a *détour* if pursued by the enemy.

The detachment should be quartered in a few houses only, near the alarm post, and part should have their horses constantly bridled. The roads in the direction towards the enemy should be barricaded, to prevent his rushing in suddenly; and, to secure the more easy and expeditious retreat for the detachment, openings should be made, here and there, through the gardens, which should be pointed out to the troops.

d) *Conduct during the Night.* The above precautions are sufficient in the day, provided the outguard possesses a commanding view of the country. But at night, other arrangements are necessary: the detachment then turns out, or is collected together in a remote barn, with their horses bridled. But if the danger is great, and the enemy near, or if you have reason to suppose he may receive information of the detachment, another station should be taken at night. As soon therefore as it is dark, the party should move to another village, or into a distant barn or garden, affording shelter from the weather; or in the summer it may remain in the open fields. No person in the village should be acquainted with the situation, and if the detachment is just arrived, it should only be known to those whose houses are occupied. *Every means should be taken to prevent the enemy from being able to find the detachment, so that it may be secure from surprise, even if he is not discovered by your patrols and outguards.*

§. 105. *Treachery.*

It is further necessary, to prevent the people from betraying the party, that, in the first place, they should be treated with as much kindness as circumstances will admit; and secondly, an order should be given for every inhabitant receiving a visit from a person belonging to another village, or purposing to quit the same, to give immediate notice thereof; and threatening that any person giving information, out of the place, even to a relation, of the station of the detachment, &c. shall be hanged as a spy; and the punishment extended to the village, which will be reduced to ashes. Thus the whole village will be made responsible for every individual, and the conduct of each inhabitant regarded with a watchful eye by the whole.

§. 106. *Placing the Outguard. Postēs d'Avertissement.*

The outguard should not be stationed at more than 1500 paces from the village, or the detachment, unless there should be at a greater distance, a high ground, from whence a more extensive view of the country may be obtained; and in this case, an intermediate post should be established, by which the alarm may be given, as soon as the firing of the outguard is heard. It is not difficult to post an outguard during the day, in such a manner, that with 2 vedettes, the country may be watched to the distance of 2 or 3000 paces. If a view of the country can be gained to this extent, the small party will always have time to mount before the enemy can arrive.

In general the country may be observed to a great distance, by a *Poste d'avertissement*, placed upon a tower or hill, particularly if provided with a telescope.

§. 107. *Patrols.*

The principal security of a post of this nature, in addition to its own vigilance and that of the outguard, depends upon its patrols. These, consisting of 3 men each, should advance 4 or 8 miles in the direction towards that part of the country occupied by the enemy. If they find that the enemy is in motion, that he is approaching, or if they fall in with his patrols, then danger may be apprehended, and the attention must be redoubled; numerous patrols must be sent out, and the position of the detachment changed frequently, and with such secrecy that no person may be acquainted with its situation. In all such cases, and whenever there is danger, the patrols must follow the rules laid down for detachments on secret marches: these rules are of the greatest importance for patrols of detachments of this nature, and of all which are weak and considerably extended. If they adhere to these rules, take care to be provided with guides, and do not proceed by the most frequented roads during the night, they will return in safety, even should they fall in with stronger patrols of the enemy. By day the patrols should be made in the woods, and at night, they should procure guides according to §. 37, and collect information in the villages. See §. 9. *b.*

In the summer, the enemy may, in this manner always be approached, when the precautions described in §. 37, &c. have been employed, and his advance, or even his intention to surprise, discovered in time. If the weather is so bad that the patrols cannot remain entirely in the open air, the men must warm themselves alternately in the nearest houses, while the others continue without to guard against a surprise; they may also, when it is necessary, sleep in this manner during the day; but it is better if they can complete their patrol, without taking rest.

§. 108. *Vedettes in the Manner of the Cossacks.*

If a post is too weak to secure itself properly by outguards, it is usual to supply this deficiency by posting double vedettes in concealed spots, from whence they have a view of the surrounding country. When these vedettes hear or see any movement, they retreat quickly and by unfrequented roads, without discovering themselves, to the nearest post; or they fire when they arrive within such a distance that a shot may be heard. Each takes a separate road, which they have previously determined upon.

It is obvious that steady, attentive men, well mounted, should be chosen for these vedettes. In winter, there should be a fire in the rear, to which they may repair alternately. In thick woods and mountains, this duty may often be best performed by dismounted men. A principal consideration is, that their situation should never be known, and that it should be frequently changed; sometimes by drawing the vedettes nearer, and sometimes by removing them to a more distant station. During the night, they should place themselves in the defiles, roads, &c. which the enemy must pass, concealing themselves behind the hedges, &c. and listening with attention. They are relieved according to the state of the weather, sometimes every 4 hours, or when it is mild, every 6 or 8 hours. The longer they remain on duty the better.

§. 109. *Guards of Peasants.*

Though little dependance is in general to be placed upon these guards, yet they may sometimes be usefully employed by parties of the above description. Orders are given to the distant villages, situated in front and on the flanks, to give immediate information of the arrival of any party or patrol of the enemy. They may also be permitted to guard some of the nearest passes, particularly in the winter; similar orders should be given to them to communicate instantly the approach of the enemy towards the points where they are stationed. If these people are well disposed, they may often be of great use. The Austrian and Prussian armies frequently availed themselves of their service. It is best when men can, for a remuneration, be procured secretly for this purpose.

§. 110. *Communications with the adjacent Posts.*

The patrols often extend to the neighbouring posts, to concert measures, and to make known what is going forward, or they communicate by the concerted signals.

§. 111. *Instances.*

Numerous instances may be cited of guards, of the above description, having been carried off, and, on the contrary, of their having maintained themselves by vigilance and circumspection.

1) In the year 1758, a post of 24 men, of the Luckner hussars, in front of Meschede, was twice carried off, with the exception of a few men; the command was afterwards given to

Lieutenant Luderitz. The post was surrounded by hills and woods, and the French received every information respecting it from the country people. Under these circumstances, the Lieutenant frequently changed his quarters, and stationed a corporal and 6 men in his rear, that the enemy might not approach unexpectedly in that direction. The vedettes were so placed, that every thing to a certain distance could be seen. Patrols were constantly sent out in front and on the flanks of the vedettes. At night, Lieutenant Luderitz mounted with his whole detachment, and caused the fire to be kept up by a peasant. The station was changed every night to some place, which had been fixed upon during the day; and the vedettes were ordered, in case of any alarm, not to retreat upon the party, but to provide for their own safety in the best manner they could.

These prudent measures, which were soon betrayed to the enemy, notwithstanding, prevented his undertaking any thing against the party. And another detachment, which did not observe the same precautions, was afterwards surprised.

2) In the year 1759, the above-mentioned Lieutenant Luderitz was posted, with 30 cavalry, near Kirchhayn, which place was occupied by the corps of Chabot. The latter had a picquet of infantry in the church-yard, without the gate, and an out-guard of dragoons more in advance, in the open country. Lieutenant Luderitz was posted with his party upon the border of a wood, which was in his rear. As soon as the French discovered him, they commenced skirmishing, but conceived that some yagers and a body of infantry were concealed in the wood, and that the hussars were only placed there, to draw them into the fire of the ambuscade. Towards evening, Lieutenant Luderitz had a fire lighted, but instead of remaining near it, he moved to another spot, which he had previously fixed on, in order to secure from a surprise. At day-light he returned to his former post, and, on the approach of evening, moved as on the preceding night. The patrols extended as far as was possible, without danger. One night about 12 o'clock, a patrol heard a movement in front of Kirchhayn, and Lieutenant Luderitz

perceiving the enemy had some design against him, gave instructions to the vedettes accordingly, and made some change in their position. He had been scarcely an hour returned, when a vedette called out, "*Who's there?*" "*Who's there?*" some shots were heard at the same time, and the enemy instantly advanced at speed to the spot, which had been occupied the preceding day, where, however, they found nothing but the fire.

Lieutenant Luderitz kept his party concealed, and every thing soon became quiet. Two patrols were sent out, who brought information that a number of parties were in the open ground, but by day-break they had all disappeared. The party had scarcely resumed its former position, before it was attacked by Lieutenant Joung, of Chabot's corps, with 30 dragoons, supported by 2 troops. They retreated as quickly as possible upon a village in the rear, where Captain von Bennigsen was posted, with a company of Stockhausen yagers, and there the vedettes, which had been dispersed, joined the party. Lieutenant Luderitz repeatedly endeavoured to advance, but was always driven back. The position, which he had quitted, was occupied by Lieutenant Joung, with 30 cavalry, who placed a corporal with 6 men in his front, in the middle of the wood. This disposition was dangerous. General von Luckner formed his regiment in 2 divisions; 4 troops were conducted by Lieutenant Luderitz through the wood, so as to come upon the flank of Chabot's dragoons, who still remained in parties upon the plain of Kirchhayn: the other half of the regiment moved directly forward. This latter party drew the whole attention of the dragoons in the wood, while Lieutenant Luderitz, with the other 4 troops, fell suddenly upon their rear, and upon the rear of the remaining parties on the plain. Forty prisoners were taken, and Lieutenant Joung died the following night of a wound, which he received. A position should never be taken in a wood, in the day time, when it may be turned, and when a situation may be chosen in the open country.

3) On the 13th of December, 1760, Lieutenant Kœnig, of Bock's dragoons was posted, with 2 serjeants and 24 privates, at Essente, a village in the neighbourhood of Stadtberg. He described his situation and operations as follows:

"In my front was Stadtberg, situated in a deep valley; between this town and my post lay a defile a mile and a half in length. Stadtberg was not occupied by either party. The left wing of Bock's corps was in the monastery of Bueren, 8 miles to my right, and 2 battalions were in Bueren. Herdehausen, 6 miles distant, was the nearest quarter to my left. Thus I had a space of 14 miles to observe, and my object was, to maintain the communication between the two above-mentioned quarters. At the distance of 2 miles in my rear, was the village of Meerhof, which was not occupied. A lieutenant of Trimback's hussars had formerly been surprised in this position, and after I quitted it, a battalion of the *Legion Britannique* shared the same fate in Stadtberg. As I occupied this post in the severest season (from the 13th of December to the 10th of January) it was impossible to remain in the open field; I therefore quartered my men in three large houses adjoining each other, on the side of Meerhof, and placed some dismounted sentries upon several heights, which commanded a view of the country: I also sent out 4 patrols of 2 men each daily, mounted on horses belonging to the country people*, for the purpose of obtaining intelligence.

The first patrol proceeded through Stadtberg, Kloster, Bredeler and Matfeld, towards Brilon, advancing thus about 8 miles: the second took the road of Canstein, in the district of Arrolsen, and sometimes entered the latter place. The French also patroled to the same place from Sachsenhausen and Waldeck Castle, being thus within 12 miles of my post. The third patrol passed through Furstenberg to Niederalmen, 6 miles distant; and the fourth to Herdhausen, where General von Mansfeld, of the Brunswickers, was stationed.

I procured provisions and forage from the places in my front; the inhabitants of the country which I occupied, having lost every thing the preceding campaign, I gave several of them provisions, for which they gratefully performed for me the duties of sentries and patrols.

---

* From this and what has been said before, it seems that the intention was to spare the horses as much as possible. The imitation of this kind of patrols and dismounted sentries is not in all cases recommended. Lieutenant Koenig was ordered to use the horses of the country people in patroling.

During the night, I seldom remained at the post after the patrols had returned; but traversed the wood silently in different directions, or in more severe weather proceeded to Meerhoff, Eisdorf or the Monastery of Dahlheim.

CHAPTER II.

Of larger Detachments, consisting of from 100 to 1000 Men, posted for the Purpose of observing a particular District, or a Corps of the Enemy.

§. 112. *Strong Detachments, consisting of several Squadrons or Battalions.*

a) Detachments of this kind form a chain of out-posts round them at the distance of 2000 or 3000 paces.

b) These advanced posts establish, in like manner, at the distance of 2 miles, smaller posts of from 6 to 12 hussars, who patrol the adjacent country, and watch the defiles.

c) In addition to which, these detachments send parties of from 8 to 15 men, to the distance of from 6 to 12 miles, or farther, both on their flanks and in the direction towards the enemy. These parties are not confined to any particular spot, and conduct themselves as if on a secret march; they are relieved every third day, and report every morning and evening to the nearest post.

§. 113.

The following Example will illustrate the above Rules.

In the year 1778, when Prince Henry's army was at Nimes, and Platen's corps near Budyn; General Sauer was detached from the Austrian army, with 200 yagers, a weaker battalion of Croats and 10 squadrons of light cavalry, towards Welwarn (Pl. I. No. VI) for the purpose of observing Platen's corps near Budyn, which was not more than two day's march distant.

The bridge at Welwarn was occupied by a company of Croats: and 1 mile beyond Welwarn, near Lautzka, a captain's outguard of cavalry was stationed, which formed a chain of posts, consisting of 80 men, and extending from the rivulet near Budenitz to Wodochod. A troop was detached on the left towards Schlan, and another on the right towards Raudenitz, both commanded by chosen officers. They were relieved every 4 days, and were directed to traverse the country in different directions (with the precautions used on secret marches) and to remain constantly mounted at night.

Besides this, a captain was stationed with 60 cavalry at Martinowes, who constantly patroled the bank of the river Eger, and reported several times every day to the outguard. By these dispositions, Sauer maintained himself during the whole campaign, in the neighbourhood of the very superior corps of Platen, without any considerable loss.

§. 114. *Further Observations respecting the Parties sent out by the above Detachment.*

The parties sent out towards Schlan and Raudenitz, conduct themselves in the same manner as the patrols, mentioned in §. 107—if they remain long in the same place, they will observe the same conduct as the detachment mentioned in Chapter I. §. 100, &c. It is, however, a principal object that they should change their stations every night, using such precautions and secrecy that no person may be acquainted with it. Several parties of or 4 men each should be sent out frequently, who will rejoin again at some appointed spot, and will conduct themselves as the small patrols in §. 107.

The parties must also endeavour to procure, from among the inhabitants, some persons who will obtain information, and will be at places fixed upon at a certain time, for the purpose of making their reports; to these places a few men only will be sent to receive the intelligence, so that the parties may incur no risk in case of treachery.

§. 115. *Stationary Detachments, consisting of* 1 *or* 2 *Squadrons, or of* 2 *or* 300 *Infantry.*

The measures requisite for the security of a detachment consisting of one squadron, or, in intersected ground, of a battalion, are the same as those laid down from §. 112. to §. 114. for stronger detachments. The only difference in the present case is that the outguards are weaker: of these only 2 or 3, of 9 men each, should be established at the distance of 2 or 3000 paces, and one should patrol about 2 or 3000 paces farther in advance. Patrols, of from 3 to 9 men each, should also be sent by the detachment, some miles in the direction towards the enemy, and upon the flanks; they conduct themselves according to §. 107.

The detachment should not remain in the same place, if the weather will allow of its moving; and when, from the severity of the season, a particular place is chosen, the party should quit it at night, or at least, should remove to some distant barn, house, &c. A picquet must always be in readiness to mount on the instant, and several sentries should be posted round the detachment.

With a detachment of this nature, much depends upon the peculiar circumstances of the case. The following account of such a detachment, posted under the command of the Captain of cavalry von Luderitz, is highly instructive.

In the month of July 1761, Luckner's regiment of hussars fell back towards Lamspringe in the territory of Hildesheim. The captain of hussars Luderitz was detached with a squadron towards Alfeld, for the purpose of observing the high road to Hanover. The regiment, to which he belonged, was about 12 miles on his left towards Einbeck; and the enemy was about 12 miles from Alfeld. Near this place, the high road to Hanover, follows the course of the Leine as far as Einbeck, so that it was necessary for him to cross the river in order to get on the road. He arrived during the night at Alfeld, and did not cross the Leine, but remained before the place, he however placed 2 outguards on the other side of the river, held a picquet of 1 officer, 2 corporals, and 18 hussars in readiness, and posted some vedettes immediately in his front; sending out at the same time some patrols on the high road towards Einbeck, who reported that they could perceive several fires on the high ground in that direction.

As soon as it was day, Luderitz, taking with him a corporal and 12 men, proceeded along the high road towards the enemy, nearly as far as the tower of Stumpfen, about 2 miles from Alfeld. Here he left the corporal with 10 men, at a sort of defile which is formed by the high ground; and penetrated, with the other 2 men, farther along the height, which is covered with trees, to the edge of the wood, from whence he could see the whole of the high road. When he had advanced about one mile and a half, he observed a patrol of 3 men of the Nassau-Saarbruck hussars, who returned after speaking to some person in the field; he also perceived that the village of Ammensen, about 2 miles from the Stumpfen tower, was occupied; and he then retired into the wood. Observing a hussar on the high road riding towards Alfeld, he discovered himself, and the man proved to be a deserter, who wished to enlist, and from whom he learnt that there were two squadrons of hussars behind Ammensen, and that farther in the rear a strong corps was encamped on the high ground.

Captain Luderitz left a corporal and 6 men, as an outguard, at the Stumpfen tower, and drew in the other 2 outguards; instead of which he established, besides that just mentioned, one on the Leine towards Grossenfreen, about one mile and a half from the Stumpfen tower, and nearly 2 miles from Alfeld: and another on this side of the Leine, about 2 miles from Alfeld, at the castle of Winzenburg. These 3 outguards, established at the distance of 2 miles, could observe every thing below them on the Leine, or that might advance on the side of the enemy.

Luderitz made himself acquainted with the country people, and obtained their friendship, which, to a detachment of this nature, is very valuable: he caused the bridges over the Leine, except that at Alfeld, to be destroyed; and thus some days passed by quietly. But the outguard at Stumpfen tower was attacked; it was supported by the picquet, and the corporal was severely wounded: the enemy also approached nearer on the side of the Leine, where Luderitz was stationed. The outguard at Winzenburg perceived posts opposite to it; and Prince Xaver reconnoitred the place. Luderitz brought up 30 men to the outguard, and flankers were sent out to skirmish with the French, who quitted Winzenburg the following night. The next morning Luderitz took a corporal and 12 men, and posted them as an outguard beyond the castle, which he then entered to speak to General Schorlemmer, who inhabited it. As he perceived the French had some design in view against him, he placed outguards again between Winzenburg and Alfeld; and drew that at Freen nearer to Alfeld, as the guard at Stumpfen tower had been removed farther from that place. About midnight he left his post, and fell back, about one mile and a half towards the village of Deensen. In the morning the French made a furious attack on the place where the squadron had been stationed, and the outguards were obliged to save themselves in the best manner they could; instructions, however, had been given for their conduct in case of such an event, and one man only was made prisoner.

The next day, every thing being quiet, Luderitz resumed his old post, and occupied a bridge about a league in his rear; the planks of which were loosened, so that he could retire to either side of the Leine, and render the bridge impassable for the enemy to follow him, or to cross to the side on which he might happen to be. He also took 2 hussars, and passed silently over the mountains to the quarter whence the enemy had advanced, where he perceived the whole camp near Gross and Klein-Ruehden.

It is to be understood, that patrols were, in the mean time, frequently sent out in every direction.

Had Luderitz posted his outguards in the usual manner, at the distance of 1500 paces, and trusted too much to his patrols, his detachment would certainly have been carried off; *but the distant outguards, the extensive reconnoissances, the intimacy which he cultivated with the people, and the constant observation of the enemy's movements, enabled him to anticipate their enterprises, and take measures to prevent a surprise.*

SECTION VI.

Conduct of Cavalry and Infantry in Cantonments, and in such Quarters as are frequently occupied, when employed on an Expedition in the Winter.

§. 116. *Explanation.*

The cases supposed here, are only when near the enemy, or at least, when within the distance of from 16 to 24 miles. When the enemy is at a considerable distance, there is generally but one main guard, or in addition, guards at the outlets, which draw a chain of sentries round the village; and detachments sufficiently advanced, in the direction towards the enemy, will give notice of his approach, time enough for all necessary measures to be taken.

CHAPTER I.

General Rules.

§. 117.

The essential point in cantonments or winter quarters, is, that the party should be assembled before the enemy can possibly enter the village; it is then easy, in the day time, to decide upon the measures necessary to be taken: if the enemy is superior, the party will retreat (unless it has positive orders to maintain the post); if, on the contrary, he is weaker, he will be opposed. But in every case, the neighbouring stations should be made acquainted with what takes place. The greatest danger to be apprehended in affairs of this nature, is at night, as it is then extremely difficult to discover the force of the enemy; but if the party is assembled before the enemy can approach, the remainder is easy, as he may be resisted with much more effect by night than by day, and succours will arrive as soon as the beacons are lighted.

Therefore it is requisite for the security of the quarter,

1st. That early information should be received of the enemy's advance.

2d. That impediments should be formed to his attack, so that time may be gained for the party to get under arms and assemble, and

3d. That an arrangement should be made for the whole to get under arms with the utmost expedition.

§. 118. *To obtain early Information of the Enemy's Advance.*

a) For this purpose outguards are employed; the manner of placing them is described in §. 62, &c. and it is only necessary to observe here, that guards, posted at the distance of 1500 paces, should be provided with *marrons*, or that intermediate posts should be established between them and the quarter, who will fire as soon as they hear the firing of the guard. Without this precaution, intimation of the enemy's advance would reach the quarter too late.

The enemy is not to be expected to advance in front only, or on the ground easiest of access. The fault of entertaining this idea was committed by the French at Wettern, in 1759, and directly afterwards at Niederweimar. All wars furnish many instances of surprises having succeeded owing to this false supposition. The rule that the enemy is more to be feared in rear than in front, and that a soldier may pass where a goat can, is not always observed, even by the best troops.

If it is found impossible, or is not deemed necessary to surround the place with outguards, a chain of sentries at 400 paces, when the party consists of infantry, or of vedettes at 600, when cavalry, should at least be drawn round it at night. Without this precaution, it might be apprehended, that the enemy could, on a favourable opportunity, rush into the place before the men could get out of the houses.

This chain is also useful, even where a chain of guards is posted at some distance; for the enemy may, by some accident, or through a neglect, succeed in surprising a guard. It is in fact, to the post, what the colour and standard guards, with their sentries, are to a camp.

b) The second mode of obtaining early information of the enemy's advance, is by patrols.

In order that the enemy may not undertake any enterprize, without the party being apprized of it, the patrols should, when he is near, cross each other's route between his quarters and their own. But when the enemy is at a considerable distance, 2 sorts of patrols should be made: the first consisting of 3 men, patrols only a few thousand paces from the outguard; they examine every thing, and proceed, particularly at night, to those points from whence the enemy may most easily advance. The second sort of patrols advance from 4 to 12 miles, according to the distance of the enemy, and are stronger or weaker, as circumstances may require, and the strength of the detachment will admit.

§. 119. *Of the Impediments that should be formed, to prevent the Enemy from entering the Place before the Troops can assemble.*

When a village is situated near a wood, an impediment may be created, by forming an abbatis on the side towards the enemy, not close to the village, but at 2 or 300 paces from it; so that, after the firing commences at the abbatis, between the guards and the enemy, some time may elapse before he can enter the village. Defiles, in the direction towards the enemy, should be rendered impracticable and guarded, by which his advance is retarded, and, in the mean time, notice of it is conveyed to the quarter. If one or more battalions are quartered in a village or small town in an open country, and the houses are close together, it may be surrounded with an abbatis, if trees can conveniently be had; when trees cannot be procured, or you are desirous to spare the wood, trous-de-loup or palisades may be used instead, and block-houses may be erected, at proper intervals; these will check the enemy until the troops can assemble at their alarm post.

It would be extremely dangerous to fix the alarm post near the abbatis, and to defend yourself in the village against the enemy, who would certainly penetrate in some part, and the whole would then be dispersed. Many instances of this have occurred.

If the village is too extensive, a part only may be inclosed.

It has been frequently proposed to construct a redoubt in villages, which cannot well be inclosed, into which the troops may retire, in case of being attacked. This is however attended with much inconvenience at night, if shelter against the wet and cold is not provided for the men within the redoubt; but, if this is done, the party will then be secure and able to maintain itself, if the work is sufficiently well constructed. The abbatis or trous-de-loup formed round the village, as well as the redoubt, require much labour, for which sufficient time is seldom allowed; and you must not neglect, at the same time, to watch the enemy's movements, while at a distance, or you will act in the same manner against the enemy, whether he is strong or weak, and will be unable to decide upon the measures necessary to be taken. An exception however exists, when ordered to maintain a post to the last: all means of defence must then be instantly resorted to, and, in that case, a good redoubt, into which you can throw yourself, is of the greatest advantage. The mode for constructing such a work, will be found in the part which treats of entrenching.

The danger which may often arise, from sufficient attention not been paid to the movements of the enemy, was experienced by a detachment of the allied army, at Nordheim, in 1760. Einbeck, Moringen, &c. were occupied by the allies, and the most advanced posts were placed at the distance of 3000 paces, in the direction from Nordheim to near Höckeln, beyond the river Leine. In this situation, the detachment was attacked about 2 o'clock in the afternoon, at Nordheim, by a superior force, which came over the mountains from Göttingen. The superiority of the enemy was not discovered, until he had arrived at the gate, when the detachment endeavoured to fall back upon the nearest post at Höckeln, but it was too late, and the greatest part were made prisoners in the retreat.

This detachment would not have been in so much danger at night, as the enemy could not then have derived so great an advantage from his superiority.

When infantry are placed in a town, which is inclosed with a ditch and walls, or otherwise fortified, two cases may occur:—

Either the party is to hold out until reinforced; or it is not.

In the first case, the place should be put in the best possible state of defence. In the second, it is only necessary to secure it against the first attack; and to leave it, under all circumstances, during the night, and also in the day, whenever the enemy appears in superior force. The fortifications are then of no use, except against a surprise, which might, perhaps, be possible, if the enemy succeeded in passing the patrols and outguards undiscovered.

If a party is ordered to maintain itself in a place, which is too extensive for the garrison, part of it only should be put in a state of defence, that is, if it is possible to separate one part from the remainder; and if that cannot be done, a strong redoubt should be erected near: but if the time, ground and other circumstances do not admit of that, the circumference of the place (perhaps the walls), should be guarded, and the men assembled in one or two points, ready to fall upon the enemy, wherever he makes his attack, and attempts to scale the walls. Had the French done this at Pattenberg, in 1760, Major Scheiter would not have succeeded in making them prisoners, with a corps not more than one third of their number. Munden might certainly have been maintained, in 1759, if the men had not been scattered over the whole extent of the walls, by which they were prevented from making any effort to repulse the enemy near the *Great House*, the place where he scaled the walls.

CHAPTER II.

Examples and Explanation of the General Rules.

§. 120. *Infantry in a Village.*

a) *The Party is ordered to fall back, when attacked by a superior Force.* In Pl. V. No. II. is a village, situated nearly the same as the town of Neustadt, which in the winter of 1757, was occupied by the French, and served as an outpost towards the allied army. It is supposed, that a battalion is posted in this village, with orders to fall back either towards E, or to the nearest quarter on the other side of the river, if attacked by a superior force. If this case is imagined, as it occurred with the French, in 1757, this battalion would, if obliged to retreat, have to fall back 12 miles in the direction of E, as far as Hanover, the nearest strong garrison.

Being posted between a morass and a river, it cannot easily be turned; but as this is not impossible, by a considerable detour, the village is inclosed with pallisades or trous-de-loup, and guard-houses are erected at A, B, and C, in the manner described in the part which treats of entrenching. Sentries are placed at some hundred paces in front of these guard-houses, and at night several additional ones are posted between them, nearer to the village. A stone building at D, likewise serves as a guard-house. The main guard is in the church-yard, on which that in the stone building D, and those in the guard-houses, at A, B, and C, are dependant; the whole are reinforced at night by the picquet, and on the first alarm, all assemble in the church-yard; and should the enemy force his way to any part, the party will quit the village by the ford, or by the way of D and E.

If an attack is made in the day, the enemy's strength may be easily discovered, from the tower, and measures taken accordingly: if he is considerably superior, you retire without loss of time. But in the night, the risk of being surprised is greater; this however is not then easy, as the enemy is not so well acquainted with the country as you are, and in the darkness may lose his way.

If a post of this nature is not provided with cavalry, to watch the country towards the enemy, it will always run great risk of being one day attacked on all sides by a superior force. But if provided with cavalry, the enemy's intention may be discovered in good time, the nearest quarter acquainted with it, and such measures taken as circumstances may render necessary.

b) Suppose this post was ordered to maintain itself until succours arrived. Redoubts should then be erected at A, B, and C, and the building D, put into a state of defence. But as one battalion would not be able effectually to defend so large an extent; one redoubt of greater strength should be erected about A, and the others should be smaller: the former should be sufficiently large for the whole battalion to retire into, in case of necessity; the others at B, C, and D, might then be simply guard-houses; but the pallisades will be necessary notwithstanding. By this arrangement, you would be secure from surprise, and be in possession of a strong redoubt, in which you might hold out, if seriously attacked. As a farther precaution, the companies might be drawn together during the night into the houses nearest to the redoubt, in which a large building, covered with earth, would be of great advantage.

§. 121. *Infantry posted in a Village near the Enemy.*

The Prussian regiment of Thadden, was, in 1778, cantoned in the village of Dittersbach, Pl. III. No. II. A battalion of Rosenbusch hussars was stationed at Pfaffendorf, 5 squadrons of the regiment of Wurtemberg at Schreibendorf, and the post of

Altweisbach was occupied by 100 cavalry, 30 of whom were detached to Neuweisbach.

The arrangements for security, adopted by Colonel Heilsberg of the regiment of Thadden, were as follows:

a) Two redoubts, one of which was erected at *d*, and the other at *c*, (neither were completed, and they were both commanded by higher hills.) One officer, 2 non-commissioned officers, 1 drummer, and 30 privates, with 2 guns, were placed in the redoubt *c*, and 1 officer, 2 non-commissioned officers, 1 drummer, and 30 privates, in the redoubt *d*

b) *Guards.* One non-commissioned officer and 15 privates were stationed at *a*, near the houses towards Pætzelsdorf, and 1 non-commissioned officer and 15 privates at the lower end of the village, towards Hasselbach, at *b*. A picquet of 1 captain, 2 lieutenants, 1 ensign, 6 non-commissioned officers, 3 drummers, and 100 privates, assembled every evening in the centre of the village, at *e*. One lieutenant and 20 men, from the picquet, reinforced the redoubt *c*, and a non-commisioned officer and 12 hussars were posted beyond *g*, towards Pætzelsdorf.

The following disposition, in case of attack, was given to the regiment, on the 17th of October.

On any alarm, the companies will assemble in front of their captains' quarters, where they will wait for orders.

The picquet will immediatly advance and place itself with its right on the wood, in a line with the village guard, and covering the road from Pætzelsdorf to Michelsdorf.

The companies are to remain in front of their captains' quarters, until the baggage is all out of the village. *(Is the baggage then the principal object?)*

In order to avoid confusion, the waggons will be formed in rear of the village, so as to be able to drive off immediately.

Should the enemy enter the village with a superior force, all the picquets and detached posts will draw together and join the captain of the picquet, *(a place for their assembly should be pointed out)*, who will receive further orders according to circumstances. *(This can seldom be expected to take place at night: every person should have received instructions for their conduct, and not be obliged to wait for orders.)*

The 1st battalion covers the redoubt on the right, and the 2d that on the left wing, *(as if the enemy's first attempt would be upon the redoubts.)*

A drummer of each company to be on the alert, &c.

On the 1st of November, to this order was added,--The batmen and waggoners will be turned out at 4 o'clock in the morning, and 1 officer and 40 men will be ordered for the escort of the baggage.

The order contained also a more minute detail for the defence of the village, redoubts, &c. as if a fortress or defile was to have been defended; whereas, the only object was to assemble the troops as quickly as possible.

It was further ordered, that the officers should pay the strictest attention to prevent the men firing without the word of command.

It will be observed, that in this instance, the following faults were committed, in regard to measures of security against surprise:

a) No outguard was posted at a sufficient distance on the side towards the mountains. The redoubts were close to the village, the upper part of which was not covered by them. The consequence was, that the enemy, advancing on the 8th through the mountains for the purpose of surprising the post, entered the village before the garrisons of the redoubts were aware of it.

Had small posts been established in the mountains, at the distance of 1500 paces from the upper end of the village, this surprise would have been impracticable.

b) A great fault was also committed in expecting the attack only on the side of Michelsdorf, and in relying entirely on the hussar outguard stationed in that direction. In the mountains was an old abbatis, but it is not very difficult to penetrate through an abbatis, even when newly formed; and if not guarded, it is good for nothing.

c) No patrols were sent into the mountains, nor even on the other sides of the village; which in this instance was absolutely necessary during the night, particularly as no guards had been posted at a sufficient distance. In this situation a few small patrols should have been constantly out, and their course so regulated that all the avenues should have been patroled every quarter of an hour.

d) The erection of the redoubts could not be of any use here; the enemy paid no attention to them, but attacked the village: had it been impossible to enter the village without passing the redoubt, the case would have been different, and they might then have been of some service; but few villages are situated so as to render this possible.

Some persons have been of opinion, that it would have been better to erect a large redoubt, into which the troops might retire in case of alarm. But even in this case, outguards should have been posted and patrols made agreeably to the instructions which have been laid down; or otherwise the troops might have been surprised in the village before they could get into the redoubt. And for what purpose was the redoubt intended? To prevent the sudden attack of the enemy during the night! it might indeed have answered this end: but it would also render any retreat extremely difficult, and it would never be adviseable to retire into it in the day time, unless ordered to maintain the post till succours could arrive.

§. 122. *Infantry in an open Town near the Enemy.*

When the Duke in the winter of 1757---8, undertook an expedition towards Zelle, by the way of Luneburg, (Pl. I. No. III.) he was obliged to fall back again upon the environs of the latter place. The French extended from Bremen along the Weser, as far as Hoya and Nienburg, and from thence in the direction of Hanover, Neustadt and Zelle. All these places, as well as Verden, were strongly garrisoned by infantry. The hussar regiment of Poleretzky was stationed in the district of Essel, between the Aller and the Leine, and a detachment of infantry in the forts of Ottersberg and Rothenburg.

About 1000 infantry, under the well-known General Chabot, were stationed at Hoya, (see this place in Pl. I. No. II.) their disposition was as follows: the suburb *d* was occupied; and two guns were placed at *d* in front of the bridge over the Weser, on the side of the enemy. Several outguards were posted at such a distance in front of the suburb, that no surprise could succeed in that quarter. Infantry guards were stationed at the outlets *a*, *b* and *c*, and the main guard was at the bridge. Patrols were sent down the Weser; and as the villages towards Bremen and Nienburg were occupied with cavalry, this arrangement appeared to promise sufficient security.

Notwithstanding which, these posts were attacked at broad day, and almost surprised before the necessary measures could be taken.

The Crown Prince crossed the Aller with a strong detachment, and approached Barmen, a village below Hoya. (Pl. I. No. IV.) Here he formed the detachment into 2 divisions, crossed the Weser with one part, and proceeded down both banks of the river: in this manner he attacked Hoya on both sides at the same time, and drove the garrison into the castle *d*, where it surrendered by capitulation.

Wherein consisted the fault of the French in this instance?

a) They supposed they could only be attacked by the suburb on the side where the allied army was posted. It must, however, be stated in excuse that they sent patrols along the Weser towards Barmen; and, indeed, 2 of these patrols passed the column of the Crown Prince without perceiving it, which was perhaps owing to the bad weather.

b) The outlets of the place at *a*, *b* and *c*, were not barricaded, it was therefore impossible to hold the enemy in check long enough to break down the bridge, and secure a retreat upon one side or other of the river.

c) No outguards were posted at 1500 or 2000 paces from the town, which would have been perfectly secure, being separated from the enemy by the Weser. Had this been done, the Crown Prince must certainly have been discovered. The security afforded by patrols is very uncertain.

d) There were no *postes d'avertissement* placed on towers or high houses, possessing an extensive view of the country.

All the faults appear to have originated in one cause;

the idea that the enemy could not get into the rear of the place.

It now remains to be considered, how the defence of this post ought to have been conducted. In the first place, What was this post relatively to the whole? It was of great importance, particularly with respect to the passage of the Weser, as the enemy, by obtaining possession of it, was enabled to place himself between the quarters, and render their junction impossible.

Thus, therefore, the principal consideration should have been the defence of the bridge, 1st. in case the enemy should advance in front, and 2d. the demolition of it, if the attack should be made in the rear. This done, the question was, whether the garrison could maintain itself in the places until the arrival of succours; or whether, confiding in its own strength, it should withdraw?

It will be immediately apparent, that the latter should have been decided upon, as, for the defence of the place, it would have been necessary to occupy an extent of 4 or 5000 paces, and the ground did not present any natural impediments to an attack. It is thus evident, that there was an absolute necessity of being informed of the enemy's approach, by means of outguards, patrols, and *postes d'avertissement*. (§. 106.)

If the enemy had then approached by the rear, the bridge should have been destroyed, and the troops withdrawn from the place. Had he advanced in front, preparations should have been made for destroying the bridge, and it should have been defended, while the country in the rear should have been scoured, in order to be as secure as possible from attack on that side.

§. 123. *A Hussar Quarter.*

When the French army occupied the quarters, as described in §. 122, the French hussar regiment moved from Bothmer on both sides of the Leine towards Neustadt, (Pl. I. No. IV.) Part of this regiment, about 2 squadrons, remained at Stœckendrebber, a village near the Leine, for the purpose of observing that neighbourhood. Neustadt, Nienburg, and Hoya were occupied by infantry, so that the hussars at Stœckendrebber might be said to have an outpost towards the enemy, and as the latter could cross the Aller at Ahlen and Rethem, and the Leine at Bothmer, a guard was posted at that place by the hussars.

In addition to this, the post had only a village guard at Stœckendrebber, (Pl. I. No. I) and at the outlet of the village vedettes were posted at *d* and *e*, and a third in the field at *a*.

The Prussian hussars got into the rear of this post, by the way of Ahlen, Gilten, and Suderbruch, and seized it. §. 57.

If the dispositions for security had, in this instance, been made according to §. 117, the surprise would have been impracticable. A small outguard would, in that case, have been placed in the wood towards Norddrebber at *c*, and another in that to the left at *b*: in addition to which, a third would, at night, have been posted obliquely towards Stœcken. If these guards had only consisted of 9 men each, yet 9 vedettes would have been sufficient to cover that extent.

Besides the guards, 2 patrols of even 3 men each, should have been sent out every day: one to have advanced by Gilten as far as Ahlen, and have returned from thence by Bothmer; the other by Suderbruch and Hœren to Frankenfeld, and returned by Rodewald.

If, in addition, patrols had been sent out at night by the picquet to Norddrebber, Suderbruch, Rodewald and down the Leine, this post would have been tolerably secure.

In these cases, your whole thoughts should not be given to security, but also to gain information of the enemy, to observe and perhaps annoy him. Then only are you perfectly secure, and then only do you completely perform your duty.

§. 124. *Infantry in a walled Place, and ordered to maintain itself there.*

The town of Nordheim, in Pl. V. No. I. is surrounded with a wall, 12 feet high, and 4 feet thick, and has a wet ditch at *a*, *r*, *s*, *p*, and *t*.

a) Three battalions are stationed in this town, with orders to defend themselves until they receive succours. The enemy is on the other side of the river Ruhme, (beyond F) towards Einbeck.

In this case, every thing depends upon the internal arrangements for security. In the first place, traverses should be formed within the gates, that is, an épaulement should be thrown up exactly opposite to, and about 30 feet within the gates, in such a manner, that waggons may pass through the gates, by the roads to the right and left, along the wales.

The gate affords security against surprise, and the traverse furnished with a gun, the means of resistance against attack.

b) A sort of fleche should afterwards be erected in front of each gate, as shewn in the plan at *a, e, d*, by which the walls, not provided with a wet ditch, will be defended. These fleches or redoubts have a strong profile, the ditches are pallisaded, and trous-de-loups placed in front of them.

c) Loopholes should be formed in the semicircular parts of the wall; they should be 7 feet from the ground, that the enemy may not be able to fire in through them. By means of these loopholes, a flanking fire will be gained upon the enemy, should he attempt an assault. But as the sides of these loopholes are apt to break away, the better way is to make a hole in the brick work, into which a piece of wood should be let, and the loophole cut through that.

d) In the day, small non-commissioned officers' guards will be detached by the gate-guard, outside the gate, to A, F, D and G, who will communicate every thing that occurs to the gate-guard, or fire, should the enemy appear unexpectedly: the gates will then be immediately closed, &c. Without this is done, you are not secure from surprise. §. 60.

At night, these guards are withdrawn, and in their stead, a chain of sentries is formed at 3 or 400 paces distance round the place, and at 2 or 300 from each other, in order that too great a number of men may not be taken up on this duty. If this were not done, the enemy might approach close to the wall, and make his observations without being discovered. The main guard is at *b*, and the gate-guard is reinforced at night by the picquet; they then post sentries at 50 paces in front of the ditches of the redoubts, and also behind the wall of the town. During the day, the main guard should station a non-commissioned officer and a few men, provided with a telescope, upon the tower, to observe the country round.

In case of alarm, one company is to repair to each gate, and half a company to each of the semicircular towers; the other 10 or 11 companies will rendezvous at the main guard.

The part which is attacked reports immediately to the main guard. The guns are in the redoubts.

e) The straw and hay should be collected in some remote place, that the town may not be entirely consumed by means of it. Water should be kept ready in the garrets of the houses for the purpose of immediately extinguishing any fire.

f) The garrison and the inhabitants should be supplied with provisions for some time; or the place may soon be compelled to surrender, when attacked.

SECTION VII.

Conduct of Detachments in search of the Enemy.

§. 125. *General Rules.*

a) For the conduct of the march, see §. 25, 26, &c.

b) You should endeavour to keep the whole, or at least, part of your detachment undiscovered, from which many advantages arise. The enemy may be attacked unexpectedly, or, if he is inferior in numbers, and has only seen a part of the detachment, he may be induced to engage, conceiving himself stronger; but if he is stronger, still there is a possibility of drawing off in safety, or at least, of gaining time for the infantry to throw itself into inclosed ground.

If the enemy has light cavalry, and you have only heavy, you may perhaps, by not being entirely discovered, avoid an action, which, under such circumstances, would be extremely disadvantageous.

c) If you discover the enemy, or receive intelligence of him, on which you can rely, and have not been yourself discovered, you may perhaps be able to draw him into an ambuscade, if he is not stationary.

For this purpose, you must place part of your detachment in some wood or barn, &c. and with the remainder approach the enemy in such a manner as to lead him to believe you do not expect to meet him; then form sometimes in one position, sometimss in another, and appear to be perfectly lost and confused; taking first one and then another direction, never approaching strait upon your ambush, but moving so as at last to bring it in the enemy's rear.

Ambuscades may also be prepared in more than one place, and the enemy thus attacked unexpectedly in different points; or a trifling one may be formed in front, and if the enemy discovers this, he will not expect another.

The principal point to be observed, with respect to an ambuscade, is, that it does not attack too soon. Lieutenant-colonel Ewald, in his treatise on the duty of light troops, remarks how frequently this is the case. The enemy should not be attacked before he has discovered you, or when you can fall upon his flanks or rear, *and when he is in such a situation that he cannot escape.*

d) If you discover the enemy, or have attained your object, you should retire directly, or follow up your advantage upon the instant.

In the one case, you are secure from surprise; and in the other, something of importance may be executed in the first moment, (see §. 59) which would afterwards be impossible.

§. 126. *A small Detachment in search of the Enemy.*

In this case, the rules for strong patrols (§. 1, &c.) must be observed; or, if the detachment is very weak, those for small patrols. (§. 107.).

You may also divide the detachment into 2 or 3 parties, even if they do not consist of more than 3 men each, and then proceed to the point of destination, by different routes; the parties keeping a distance of 2 miles or more from each other.

Each party will observe the same precautions as small patrols (§. 107.) A place should be pointed out, to which the parties are to fall back, on discovering the enemy: and, as each should communicate to the others, according to circumstances, any intelligence respecting the enemy, the points, at which each party will halt at certain periods, should be previously fixed upon. It being necessary that the parties should remain concealed, the man who conveys the intelligence should fire a shot, when he arrives at the spot fixed on, which will be answered by the party; and by this means, each will be able to find the other. In order that this intelligence may not be discovered, the man will be furnished with some false documents, which each commandant will be aware of.

It is to be understood, that a good map is necessary on these occasions, which may be corrected.

§. 127. *Of a large Detachment patroling a Country, for the purpose of destroying the Enemy's Magazines, or dislodging his Parties.*

a) If the enemy is supposed to be at a certain place, the march should be directed thither, with all the precautions necessary upon secret marches; it may thus be possible to surprise him, in executing which, the observations in the section treating of surprises, must be attended to.

b) If there is reason to believe that the enemy has magazines somewhere in the country, or is stationed there, in numbers not too considerable, parties should be detached to some of the places suspected; the points at which they may fall in with the main detachment, or where they may convey intelligence to at fixed periods, must be determined upon. The rules laid down in Section I. Chapter II. are to be observed on the march. If the smaller detachments are weak, a surprise may perhaps be effected; but if you have reason to fear the enemy is superior in number, it will be necessary to act as if on a secret march.

When Colonel Freytag, in 1762, was directed, with a detachment from Einbeck, to destroy the enemy's magazines upon the Werra and Fulde, he proceeded into that country by forced marches, and detached 200 cavalry to the Fulde, while he moved with the main body along the Werra: every thing was so arranged, that the detached party might fall back upon the main body, without danger of being cut off; both these sent out smaller parties on their flanks, whenever it was found necessary, which were also enabled to retire upon them. Thus the whole was divided into two main bodies, connected with each other, and those branched out again into smaller parties, and the strength was preserved, while the advantage of acting in separate bodies was gained.

Colonel Freytag marched from Einbeck, by the way of Katlenburg and Radolfshausen, as far as the country between Witzenhausen and Allendorf.

On the evening of the 19th, near Allendorf, he detached Captain Campe with 200 cavalry towards the Fulde and on the 20th proceeded with the main body along the Werra towards Wanfried, where it was supposed the enemy had some magazines. Captain Campe marched all night, and in the morning, reached the neighbourhood of Rothenburg, on the Fulde; here he learned that a convoy of 100 loaded waggons was on the road to Melsungen. He detached Lieutenant Thiele with 40 men to seize this convoy, while he himself proceeded with the remainder to destroy a magazine at Rothenburg. Lieutenant Thiele accomplished his object, but was obliged by a detachment of the enemy's dragoons, which came from Melsungen, to fall back upon Captain Campe.

Several officers and men, belonging to the enemy, were made prisoners at Rothenburg; from whence Captain Campe continued to move along the Fulde, and detached Captain Engel, with 100 men, to Hirschfeld: but this place being occupied by 200 infantry, Captain Engel turned his attention to the magazine without the gate; the picquet, which guarded it, consisting of 30 infantry, was attacked by Lieutenant Scheiter; 5 men were sabred and 12 taken; the magazine, which was considerable, was seized. Captain Engel then fell back upon Captain Campe, and the latter, on the 21st, also retired by Eschwege, upon the main body, which was at Wanfried, and had found a small magazine at Eschwege, but had in vain attempted to discover the enemy on the Werra. The whole detachment had returned to Duderstadt by the 22d, having marched nearly 16 miles every day. Expedition is a principal object in undertakings of this nature.

§. 128. *A small Detachment is ordered to observe the Enemy on his March.*

The commander of a corps, or of an army, receives information that a detachment of the enemy is in motion; a small party of cavalry is in consequence sent out to ascertain the truth of this intelligence, and the direction which the enemy has taken.

General Rules.

a) The march should be executed in the manner pointed out for *secret marches*. §. 37, &c.

b) Information respecting the enemy will be obtained

by interrogating the inhabitants, and causing them to make enquiries, sooner than by any other means. Having discovered by enquiry where the enemy is, he should be followed from place to place, and upon arriving at any place which he has very lately quitted, guides should be procured to lead the party secretly close to the road, by which he must pass, in order to procure a view of his detachment. Previous however to this, one man should be sent back with the information that has already been obtained. By means of a few ducats, persons may always be gained on such occasions, who will prove able guides, and will be useful in other respects; they may also assist in discovering the enemy.

c) When approaching near the enemy, a small reserve should be left near some defile; that, if discovered unexpectedly and pursued, the enemy may be checked by it for a short time.

d) If pressed hard by the enemy, after having discovered him, the party should disperse by different roads, previously pointed out.

Instance. When the Crown Prince, in 1759, marched from the camp near Crofdorf to Fulda, in order to dislodge the corps of Wirtembergers, he was informed that 150 hussars had passed his left near Hopfgarten. Lieutenant Lange of Bock's dragoons, was detached, with 2 non-commissioned officers, 24 dragoons, and 8 hussars, to observe, and bring positive information respecting this detachment. At the village of Hopfgarten, Lieutenant Lange was informed that the enemy had taken a guide thence and proceeded towards Alsfeld; he therefore also procured a guide there, and advanced slowly, causing the return of the enemy's guide to be watched, who was fortunately met with, and who stated that the enemy had marched to Alsfeld. This information was instantly forwarded to the Crown Prince by a hussar. At Schwarz, the first village on the road to Alsfeld, Lieutenant Lange found, that the enemy had fed their horses there only a quarter of an hour before; he therefore promised a florin to a peasant, to conduct him through the wood, so that from the border of it he might be able to see the enemy's detachment upon the road to Alsfeld; this was done. The party proceeded by unfrequented roads, Lieutenant Lange being about 50 paces in front with the guide, and the men following in file. The enemy's hussars were on the low ground, in front of the outlet of the village, half a mile on this side Werges, beyond which place they had a vedette; and at the distance of about one mile and a half, Lieutenant Lange perceived some of his own corps moving.

Lange formed his detachment in 3 divisions, one of which advanced in front, being in single rank, and the men extended at 50 paces from each other; he thus approached rapidly against the dismounted hussars, who immediately threw themselves on their horses, and took to flight. The enemy now saw our corps in their front, and Lange's detachment in their rear.

Lange incurred no risk in this affair; particularly, as a ravine prevented any further attack, either by him or by the hussars. He took 17 waggons at Werges.

SECTION VIII.

Conduct of a Detachment in reconnoitring.

§. 129. *Of reconnoitring an Enemy's Post with a few Men.*

The same conduct is to be observed on the march, as is prescribed for secret marches, (§. 37) and on other points. See §. 45.

§. 130. *Of reconnoitring the Enemy with a strong Detachment.*

a) If a part of the detachment consists of infantry, they should be left behind near the defiles, and in the intersected parts of the country, through which the enemy must pass in pursuing.

b) When approaching near the enemy, reserves should be placed in concealed spots to the right and left, which will cover the retreat of the detachment, and attack the enemy in flank, should he pursue: on perceiving them the enemy will proceed with caution, as he will not be aware of their strength; their horses being fresher, they may also skirmish with him, and gain time for the party to retreat.

If the enemy is strong, particularly in cavalry, the principal part of the detachment may be left behind in reserve, divided into small parties, within the distance of from half-a-mile to 4 miles; by which means, when pursued by the enemy, continual reinforcements will be received.

It may be a good plan to prepare an ambuscade with this part of the detachment, and endeavour to lure the enemy into it, by small parties; but this depends entirely upon the country being favourable, and seldom succeeds.

c) The part of the detachment, destined for the reconnoissance, as soon as it is discovered, attacks the enemy's outguards, and drives them in as fast as possible: one part then extends and skirmishes, with a view of making a few prisoners; while the engineer or officer, charged with that duty, makes his reconnoissance, approaching as near to the post as circumstances will permit.

d) If the enemy consists of cavalry, and is able to turn out in time, every endeavour must be made to complete the business as soon as possible, and then fall back, without engaging more than is necessary. The object in view is not to beat the enemy.

e) In the manual for cavalry officers, a rule is given to alarm the enemy in different points at the same time. By this he is obliged to discover his whole strength, and is prevented from adopting the necessary measures; it also facilitates the escape of at least a part of the detachment.

Example. Suppose Weidmansdorf is to be reconnoitred from Freyberg, with 1 squadron and 200 infantry. Thirty cavalry should be sent by Brand, with orders to leave a non-commissioned officer and a few men at that place, for the purpose of watching the right flank; and the remainder should proceed through the wood, leaving Mœnchfrey on the right, and posting a few men as a reserve on the road from Freyberg to

Mœnchfrey, and should then approach Weidmansdorf, on the side of the latter place.

The main body of the detachment should march by Bertelsdorf, in which place the infantry, after some parties of it have been stationed at the passages across the Mulde, should remain, for the purpose of observing the roads which pass through it from Weidmansdorf to Mœnchfrey.

The remaining 120 cavalry should approach Weidmansdorf in 3 parties, and drive in the outguards in front of the village; while the detachment, which had moved by Brand, would advance on the side of Mœnchfrey.

§. 131. *Additional Remarks on reconnoitring.*

A reconnoissance at a considerable distance, or in an open country, requires a strong detachment, which should, if possible, consist of hussars or dragoons. Infantry are in great danger, when the enemy has been able to turn out, of being cut to pieces in its retreat, by a superior cavalry. This was very near being the case with the infantry of the detachment which Major Köler commanded when reconnoitring Burkersdorf, in 1778. Infantry, when it does form part of a detachment of this nature, must not advance beyond the intersected country.

In reconnoitring an open country, with 300 hussars and 300 dragoons, they should, according to the opinion of Generals Turpin and Warnery, be divided into 12 parties. On arriving within 3 miles of the enemy, 2 parties of dragoons should be left behind, one posting itself to the right and one to the left, examining and observing the whole country; when within 2 miles, 2 more parties of dragoons should be left, posted in like manner: and on approaching still nearer (at about one mile) the 2 last parties of the dragoons should halt.

Each of the 4 last-mentioned parties will observe the same conduct as the 2 first; but instead of extending, like them, from one to one and half miles to the right and left, the third and fourth will keep at about 1 mile from each other, and the 2 last posted will not exceed half a mile at the utmost: but this depends much on the nature of the country, as they must all post themselves behind hedges, houses, &c. and from thence observe the parties in their front nearer to the enemy. Thus the 6 parties of dragoons form a sort of wedge towards the enemy.

Two parties of the hussars then form the advanced guard, and, extending from 1000 to 1500 paces from each other, search the whole country. The 4 remaining parties of hussars follow at about 1500 paces distance; and on falling in with the enemy, the 1st and 2d parties skirmish, and the commanding officer advances with the 3d and 4th, while the 5th and 6th remain in reserve, for the support of the 4 first, if they should be obliged to fall back. Should the enemy continue to pursue the hussars, the dragoons advance successively and unexpectedly.

If the ground is so much intersected, that only a few places occur where cavalry can act, the dragoons will remain in rear in the open country, and the infantry and hussars will advance. The disposition will be generally as above, and if the infantry is pursued, the cavalry will cover its retreat over the open spots.

In the reconnoissance itself, the infantry and hussars will support each other alternately, as the ground will admit.

SECTION IX.

Conduct of Detachments when engaged.

CHAPTER I.

General Rules to be observed by the Commanding Officer, before engaging the Enemy.

§. 132.

1) A detachment, when near the enemy, or when expecting to be attacked, should keep itself as much concealed as possible; the situation where it is posted, should be so chosen, and its marches so directed, that it may not be discovered, or at least, that only a part of it may be seen. From this the following advantages will be derived.

a) If you are not seen by the enemy, you may attack and rout him, before he can put himself in a posture of resistance. If the detachment is very small, an opportunity may perhaps be gained of cutting him off entirely, or of surprising him while feeding his horses, or on his entering a place. If the enemy has been able to discover only a small part of your detachment, it may be possible to prepare an ambuscade, and draw him into it by means of the party which he has seen.

b) Should the enemy be stronger than you, it will

be necessary to withdraw, if possible, undiscovered, or to form the detachment in single ranks, and divide it into several parties, so as to give the appearance of a greater force.

c) If your detachment consists of light and the enemy's of heavy cavalry, you must engage at all events, provided your horses are not too much fatigued; but when the reverse is the case, you must retire, without loss of time, as your heavy cavalry would be unable to effect any thing against the enemy's light.

2) If you are informed that the enemy purposes to attack your detachment, an ambuscade should be prepared upon the road, which he must pass to execute his design, and a small party left to draw him more readily into the snare. This rule should be observed also when expecting an attack from a distant point.

3) If there is great probability of your being attacked, and you cannot prepare an ambuscade, then attack the enemy the moment he prepares to attack you, or even before. Experience proves that the attacking party has great advantages over its adversary; and only one exception exists to this rule,---when placed in a fortified or other strong post.

In short, when it is evident that you must engage, animate your men by addressing them briefly, but with spirit. If the regiment, to which they belong, has never been beaten, remind them of it, and avail yourself particularly of their prejudices respecting their superiority over the enemy's troops. But above all, every officer and also every non-commissioned officer, commanding a small detachment, should be perfectly acquainted with his duty in all respects, and with what is necessary to be done under particular circumstances; as for instance, that the squadron is to divide itself, on the word of command, if the enemy shews an intention of attacking on both flanks, &c.

CHAPTER II.

Conduct of a Detachment, ordered to skirmish with the Enemy, when composing Part of the Rear Guard, or when pursuing, &c.

§. 133. *Division.*

If 30 men are ordered to hold in check the enemy's skirmishers, they must be divided into 2 parties, each of which is to be commanded by a non-commissioned officer; each party is divided into 2 or 3 sections, which skirmish alternately; each section is again subdivided by two's, and each 2 men keep together and support each other.

§. 134. *Conduct of Cavalry Skirmishers.*

a) *Distance and Conduct of the Parties.* The parties are formed at intervals of from 2 to 400 paces, and mutually support each other. When retreating, one retires while the other halts, which is done alternately. If one party skirmishes with the enemy, either halted or retiring, the other forms so as to cover its retreat.

b) *Commanders of the Skirmishers.* The officer or non-commissioned officer who commands the party, remains with it; the second in command, is in rear of the skirmishers, and advances or retires with them, according to circumstances.

That the skirmishers may be more completely under command, they are ordered to keep their eyes constantly upon the commanding officer, and observe his movements, to advance and retire when he does, and never to be above 200 paces distant from him.

c) *Conduct of the Skirmishers.* The skirmishers move as nearly as possible in a straight line.

The two men, who support each other, remain together; when one approaches the enemy, the other keeps at such a distance, as to be able to support him, as soon as he begins to skirmish. The first man advances within pistol shot of the enemy, and turns himself so as to present his right side to his opponent. He should feign as if he were going to fire, and when his adversary fires, should bend forward upon his horse's neck. When within 25 paces of the enemy, he fires, and immediately prepares to draw his sword; and if the enemy is unhurt, the second comes up and takes his place while he loads. The pistols should be loaded with 2 balls each, cut in two pieces.

In skirmishing no man must remain stationary a moment: if one of the enemy's skirmishers does so, he should be fired at immediately by a good marksman; if cavalry, with a carbine.

§. 135. *Conduct of Infantry Skirmishers.*

If a detachment of infantry is ordered to clear a wood, or retires through a wood, in presence of the enemy, and an officer and 40 men are appointed as a rear guard, the following rules will be observed:

a) *Division.* The 40 men are formed in two parties, and 10 men from each party are ordered to skirmish in front.

b) *Conduct of the Skirmishers.* The skirmishers form in one line, at not more than 200 paces distance from the party. In this case also, every 2 men act together, and support each other; the first, however, does not approach nearer to his opponent than about 200 paces; as soon as he has fired, the second advances and makes ready, but does not fire until the first has loaded.

The skirmishers load behind trees or hillocks, in ditches or hollow places; they, however, always keep towards the flanks of the party.

c) *Conduct of the Parties.* The parties incline towards the flanks of the detachment, keeping about 200 paces from it, and must be very careful that they are not cut off.

§. 136. *Retreat of the Enemy.*

If the enemy retreats, the skirmishers push forward, but continue on the flanks of the parties, so as to give them the opportunity of firing with their carbines, at the distance of 150 or 200 paces.

Sometimes several skirmishers fall together upon the nearest of the enemy's skirmishers, and make a feint of attacking the party.

In executing this, great care must be taken not to advance too far from the party, that they remain in one chain, and are not cut off. If the enemy passes a defile, some will attack the rear on all sides, and others will turn the defile and threaten those that have already passed, &c.

§. 137. *Retreat of the Detachment.*

a) If the skirmishers are pursued in retreating, one fires while the other gallops back and loads; the former then falls back in the same manner, while the latter fires.

b) The same is observed by the two parties, one halts and fronts, while the other retires, and thus they retreat alternately.

c) If the enemy pushes on rapidly, the skirmishers incline outwards, and fall upon his flanks, when he attacks the parties.

§. 138. *Feints and Snares.*

a) It must be conceived, that in skirmishing, the enemy will endeavour to entrap and deceive you; will make feigned retreats, and attack your flanks from hollow grounds, or some covered spot, with a part of his troops, which have been concealed.

b) You should also endeavour to do the same by him, for the purpose of making prisoners.

It is frequently a very principal object to make prisoners, as by that means, information is obtained of the enemy.

§. 139. *Instruction.*

Without having been properly instructed, troops cannot be expected to skirmish well, as every individual depends in a great measure upon himself. In the course of the instruction, an enemy should always be represented; thus the men should attack and fire at each other, and should engage with foils. They should also be trained to the management of their horses, so as to be able, by a circular movement, to keep their adversary always on the right hand. Care must be taken, that in the exercise for instruction, the skirmishers are accustomed to pay attention to, and obey the signs and motions of their leaders; and that they support and relieve each properly. Bad skirmishers always suffer, while, at the same time, they encourage the enemy, and discourage their own party.

CHAPTER III.

Conduct of Cavalry in Action.

§. 140. *Light Cavalry engaged with Heavy.*

a) *Principle.* Light cavalry, except when a decisive blow is necessary, should never make a regular attack upon heavy cavalry, but should endeavour to annoy them by skirmishing and firing; and should harrass their rear and flanks with small parties, until they have succeeded in throwing some part into confusion, when they should instantly attack with the greatest impetuosity.

b) *Division.* If the party consists of one squadron only, it is formed in 2 divisions, which act separately, and each has its own skirmishers; if of several squadrons, they are not divided, but each has its own skirmishers; the eighth of each squadron may be formed as skirmishers.

c) *Attack*. The skirmishers approach the enemy, fire, retreat at full speed, and re-load. The squadrons should not, in general, be at more than 300 paces from the enemy, and should, together with the skirmishers, keep, as much as possible, upon his flanks. The skirmishers, upon a certain signal, assemble and fall upon the enemy's skirmishers, or upon his flanks. This should particularly be done, when the enemy gets into bad ground, where the heavy horses cannot act well; when he is making a movement, or falls into confusion; or if he is passing a defile, and is separated from his rear. The signal may be a shout, begun by the officer, and repeated by the men.

The squadrons should also attack, upon a signal by the trumpet, but not until the enemy's ranks are actually broken by the skirmishers.

The enemy must never be attacked in a defile, or in any situation, where it is not possible to get upon his flanks; as, being more capable of acting in close ranks, he could immediately break and overthrow the light cavalry.

If the enemy approaches to attack in close order, he should be avoided. The divisions or squadrons should fall back separately, and the skirmishers hover on his flanks. When he halts, they should again approach him as before.

§. 141. *Heavy Cavalry engaged with Light.*

Principle. Heavy cavalry must endeavour, above all things, to preserve order in their ranks; if that is lost, they are beat. They should cover the flanks, as much as possible, as the light cavalry, which can fall back rapidly, can approach them without much danger. They must not skirmish too much, but must repulse the enemy by their fire.

Nothing is of more importance to them than to get into narrow ground, where the flanks may be secure, there the light cavalry must be held in check by their fire; the defiles, if possible, should be rendered impracticable; and, at least, the worst mounted men (of whose escape there is otherwise little hope) sent to the rear, where they may be able, at some defile, to cover the retreat of the remainder of the party.

b) *Division and Disposition.* The best disposition for a squadron in open ground, is to detach a party on each flank; and, in case the enemy should threaten the rear, to make the rear rank face about. When the attack is made in front, as soon as the enemy's skirmishers approach near, the front rank remains firm, while the rear rank advances to the front by parties of 4 men each, alternately; these parties gallop to the front, fire, and immediately fall back, when they are succeeded in like manner by others: 16 men should be thus constantly employed, 8 of whom advance and fire, and 8 are moving forward to succeed them.

If the enemy's parties approach close, you should fire by divisions of 4 or 6 men each, (without quitting their ranks) alternately from each flank; but if the enemy advances to attack, he should be charged with the utmost impetuosity, and if his ranks are broken, you must reform as quickly as possible. If you are superior to the enemy in numbers, and that the retreat is made upon a narrow road, you may, perhaps, in this manner retire in good order, but if not, you can have little hope of safety. Every thing, in this case, depends upon the gallantry of the troops, and the manner in which they are conducted. The men are here supposed not to have been much trained to act as light cavalry.

§. 142. *Dispositions for Attacks in general.*

a) *Of Attacks in general.* The squadron, which comes in contact with its adversary, with the greatest force and impetuosity, will probably throw him into confusion. The men in general, when drawing near the enemy, check their horses insensibly, and before reaching him, several are left behind: this is not so easily done, when charging impetuously and at speed. Care, however, must be taken not to distress the horses too much, before they reach the point of attack: they may begin to gallop at about 300 paces distance, and when within 100, charge. The men should be told that an impetuous attack will be successful, but that if obliged to retreat they are lost.

b) *The Enemy may be Outflanked* by your inclining to one flank, or advancing in an oblique direction; this, however, should not be done until you begin to gallop, or the enemy may be able to make a counter-movement. If your object is to incline to your right in order to outflank the enemy's left; your right wing should direct itself upon the enemy's left, while it proceeds at a trot, but should not incline to the right until it begins to gallop. They must not however gallop too far in the oblique direction, as it is apt to create disorder in the ranks. But, as in doing this, your line will be removed more to the right, your own left wing may be outflanked; one division of the left wing should therefore advance direct upon the enemy's right; being thus separated from the others.

The 3 divisions, which incline to their right, are marked *a*, and that which advances direct upon the enemy's right *b*.

Enemy.

b a

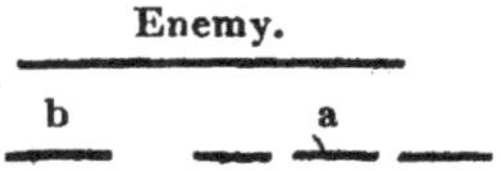

If the party consists of 4 squadrons, 3 incline to the right like the 3 divisions marked *a*, and 1 advances upon the enemy's right, bearing rather to its own left, as *b*. When, however, the party consists of 4 squadrons, it is better to form them into six divisions; 4 of which are in line, while the 5th and 6th cover the flank divisions in column, and advance unexpectedly during the attack and fall upon the enemy's flanks; the other divisions can then advance direct upon the enemy, and their shock is not impeded by moving obliquely. This mode of outflanking an enemy is often used in the Prussian manœuvres.

Warnery, with a squadron of light cavalry, defeated one of heavy in the following manner. He formed 70 men two deep; and in front of these he drew up 3 divisions in single rank and in one line; the centre division consisting of 20 men, and those on the flanks of 15 each.

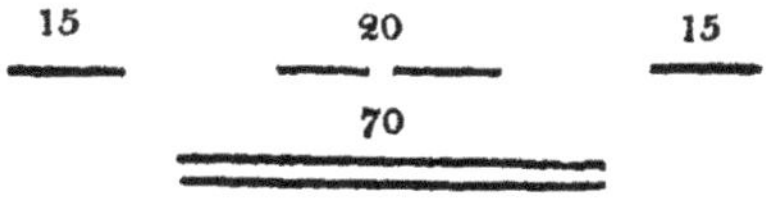

15 20 15

70

The foremost parties, which are formed in close single rank, break the enemy's order; and the 70 men in the rear complete the overthrow of such parts as may preserve their ranks.

When you have several squadrons, the intervals between them should be covered by parties of from 9 to 15 men in single rank, placed 10 paces in the rear of each other, which advance and fill up the intervals, when near the enemy. Reserves in rear of the centre of squadrons are of no use.

c) *After the Charge.* If the enemy is routed, a quarter or one-third of each squadron, previously fixed on, must pursue them and endeavour to make prisoners, this must be executed with rapidity that the enemy may not have time to rally; the squadrons follow the skirmishers. It is necessary to use great caution to avoid being drawn into an ambuscade or under the fire of infantry or artillery: non-commissioned officers should therefore be sent out to search any covered spots upon the flanks. A great object is, to form again as soon as possible; for this purpose the trumpet sounds to rally, while the squadron moves forward that the men may fall in by degrees. If the party is itself repulsed and broken, the officer should endeavour to overtake and rally those men, who have retreated farthest, and bring them again to meet the enemy.

§. 143. *Conduct of Cavalry when attacking Infantry.*

a) *Principle.* Infantry should never be attacked in front, unless it has spent its ammunition or is in confusion. This rule has exceptions, but they are very few indeed: for if infantry be resolute, cavalry will be able to effect nothing by attacking it in front, as long as it preserves its ranks; and does not fire at too great a distance. Therefore cavalry should watch the moment, when confusion takes place, or when they may be able to get upon the flanks of the infantry, &c. and should penetrate successively into the part of the line, which is in disorder.

b) *Attack of Infantry.* But should it be necessary for cavalry to attack a body of infantry, which are in close order; it should first endeavour to induce the infantry to fire, by causing small parties to advance to within about 300 paces in front, and then attack one point *en échellon.*

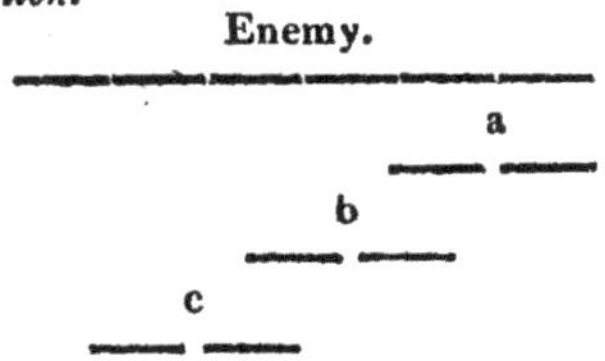

Enemy.

a

b

c

The échellon *a* consists of one squadron, and, at the distance of 150 paces, is followed by the échellon *b*, one third of the right of which covers one third of the left of the preceding échellon. The small bodies engage the enemy and the echellons endeavour to pierce through their ranks; if the first does not succeed, perhaps one of the following may. If the enemy has guns, the cavalry will begin to gallop when within 600 paces, otherwise when at 400 paces distance.

c) *Attack of a Square.* The skirmishers must endeavour to induce the enemy to fire, and the square must be constantly harrassed if it retreats; should any disorder take place in its movements, or any openings present themselves, that part must be immediately attacked. But if the enemy does not move, and will not waste his ammunition upon the skirmishers or small parties; one of the angles of the square must be attacked en échellon, if an attack is absolutely necessary.

If a detachment of cavalry pursues a detachment of infantry, much inferior to it in numbers; the best mode of attack is, (provided the enemy is not supported) to cause a part of the cavalry to dismount, to approach an angle of the square and commence firing, which the enemy is then also obliged to do: by this means the men will become less cautious, will fire at random, exhaust their ammunition and injure their flints. This dismounted party should not approach nearer than 300 paces, and should be formed in single rank; it will not then suffer much. The mounted part should remain near at hand, and watch the moment when this firing creates confusion, or the enemy begins to move without preserving proper order. It is surprising that cavalry have so frequently been sacrificed to no purpose, by repeated attacks upon infantry; when the object might so easily have been obtained by dismounting a small party.

The Hanoverian regiment of dragoons of Estorf, during the seven years' war engaged the enemy's infantry three times successfully, having a part of their men dismounted.

§. 144. *Further Rules for the Attack of Infantry.*

a) *The Time of Attack.* The rule, that infantry should not be attacked until it has fired some time, confusion has taken place or the order of its files is lost, is of the greatest importance; for if the cavalry do not succeed in breaking them in the first instance, nothing further can be hoped: the former are encouraged, and the latter dispirited.

b) *Place of the Attack.* Here the great point is, that the infantry should be upon a plain, and that it should never be engaged, while in intersected ground. The cavalry therefore should remain concealed, in order to induce the infantry to enter the open country with confidence.

c) *Means by which Infantry may be induced to commence an irregular Fire.* The skirmishers, if the infantry will not fire at them at a distance, should gallop up closer, discharge their carbines, retire at full speed, and reload. If this does not succeed, the same should be performed by small bodies; and if the infantry once fire, it must be returned at intervals, while the parties retire, imperceptibly, to 400 paces distance, and hover upon the angles of the square, then advance suddenly against one side of the square, and retire again with rapidity, &c. By this means it will probably be at length brought to commence an irregular fire, and throw away its ammunition.

d) *Infantry should not be attacked, except under particular Circumstances.* Cavalry should never engage infantry, unless they are much superior in strength, or have orders to that effect. This, however, refers only to a detachment of infantry of above 100 men; more is to be hoped from the attack of smaller bodies in open ground. A great deal however depends on the skill and bravery of the troops on either side.

More may be hazarded against light infantry, which are in general less disciplined, than against brave and steady regular troops. If infantry has already been beaten, and is on its retreat, success from an attack upon it may more reasonably be expected, than under other circumstances.

§. 145. *When the Infantry is accompanied by Cavalry.*

In this case, attack the cavalry first, and the infantry, in the mean time, must be harrassed by a few skirmishers. If the cavalry are beaten, you have then only to contend with the infantry; and if this is too strong to be engaged or attacked with a hope of success, still something has been effected by the defeat of the cavalry.

This rule, though so simple, is notwithstanding frequently neglected; as was the case with the Austrians, who, in 1778, pursued the Prussian detachment under Major Kœler, from Burkersdorf as far as Trautenau.

CHAPTER IV.

Conduct of Infantry in Action.

§. 146. *Examination of the Muskets.*

The muskets, and particularly the locks, should be carefully examined, previous to an action, and great attention paid to the flints, both as to whether they are good and properly secured, &c. A formal inspection is useless on such an occasion; but the men should be told that every thing depends on their taking good aim, and not firing at too great a distance.

§. 147. *Movements.*

Nothing is so dangerous for infantry as to lose the

order of their files; if this happens in presence of the enemy, all will fall into confusion, and panic probably ensue: therefore, movements immediately in front of the enemy, should not be performed in too much hurry; it is from this circumstance, that most squares have been defeated, as was the case with that formed by Fouquet near Landshut.

§. 148. *Firing.*

It is a general rule to fire only at a short distance.

At	150 paces,	every	2d	Ball will strike a battalion,
	200 ——		5th	
	300 ——		7th	
	400 ——		15th	

provided the whole level carefully; but as this is seldom the case, much fewer balls comparatively take effect at long distances. In addition to this, if the men have fired for some time at a long distance, they lose their coolness and attention when the enemy approaches near, discharge their pieces at random, and confuse their files; the musquets become dirty and clogged, and the flints blunt and greasy; the rear ranks open too much and level high. The smoke will also conceal the enemy; and the men having commenced firing at too great a distance, lose confidence on perceiving their fire has no effect; imagining the fault to be in the general inefficacy of fire, and not in the distance. Thus while our men lose confidence in their fire, the enemy learns to disregard it, and becomes encouraged. The Prussians lost the battle of Jägerndorf by commencing their fire at too great a distance.

§. 149. *Fire against Cavalry.*

When a battalion is attacked by cavalry on its whole front, a volley should be fired when the enemy is within 40 paces. The pieces should be loaded each with 2 balls, by which a double effect will be produced at from 40 to 60 paces; as at that distance the balls will spread from half to one and half feet, and yet retain force sufficient to kill a man. Three Hanoverian battalions in this manner defeated the French cavalry, which had surrounded them, near Crefeld.

The general discharge of a battalion, produces an instantaneous and decisive effect. If 100 men are killed, out of several squadrons, by 1 round, fired by 8 platoons, the cavalry will still preserve their order; but if the same number are killed by one volley, it is most probable they will immediately fly. Besides which, in firing by files or platoons, those which fire first, must commence when the enemy is at a great distance, so as to allow the succeeding ones to fire when he is at a proper distance. Thus the greatest effect of the fire of the battalion is not obtained, the men fall immediately afterwards into an irregular fire, and are then exposed to the disadvantages pointed out in the preceding §.

In firing at 200 paces distance, every 5th ball only will take effect against cavalry, but at 40 paces, the pieces being loaded with two balls, every one will do execution. But if you are desirous of firing twice, it will, in the first round, be impossible for want of time, to load with two balls; and in the second the men will not be sufficiently cool and careful in the loading and firing; thus so great an effect will not be obtained, and perhaps the second fire may be prevented by the rapid advance of the enemy.

If the enemy attacks *en échiquier* or *en échellon*, it will be necessary to have recourse to platoon firing. But only the two platoons nearest the point of attack, should fire, and that alternately, that is, one fires while the other loads: the first will fire when the enemy is about 50 paces distant, should he fall back, the second will reserve its fire; but if he continues to advance, it will fire when he approaches within about 20 paces; there is no doubt that this will compel him to retreat, and the first platoon will have reloaded and will be prepared for a second attack.

By the common file and platoon firing, the battalion is drawn into an irregular fire, which leads to confusion. The men do not fire at the proper distances, nor at the points where the attack is made.

§. 150. *Conduct of a Corps of Infantry surrounded by Cavalry on a Plain.*

If the Flanks are not covered, a square should be formed as soon as possible.

A battalion forms one square. Several battalions form each a square, and the line then retires by alternate squares.

The squares *bb.* retire 150 paces behind *aa.* which fire and fall back the same distance behind *bb*, &c.

In the prussian service, when 3 battalions are together, it is customary for that in the centre to remain in line, while the others form squares upon its flanks; by this a greater fire is gained to the front than if 3 squares were formed.

Count Schaumburg with 800 men, formed 4 squares, which he placed according to the following figure.

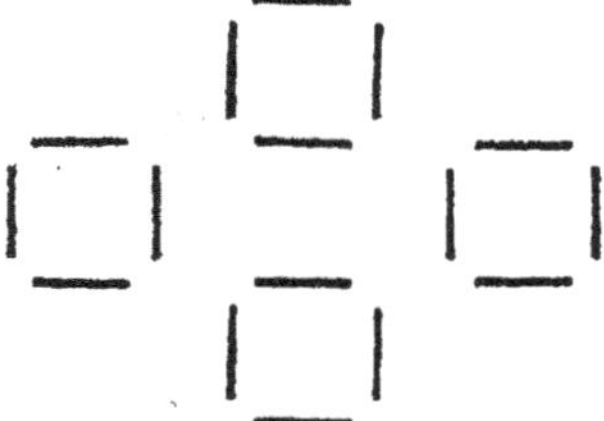

The angles of the squares did not join, on account of the firing.

That 200 infantry, formed in a square, may resist the attacks of any cavalry, has been proved by the Swedes near Tarnow in 1758: the Prussian cavalry superior in number could obtain no advantage over them. There are however many instances, where cavalry have broken into squares; among others that formed by the infantry of Scheiters corps, near Dorste in 1760; and again so lately as 1778, the square formed by the detachment sent from Glatz to support the blockhouse at Schwedelsdorf. But how did that happen? Why, the men fired as they chose, without word of command, and in consequence lost the order of their files; they got *clubbed* and run this way and that, as the enemy advanced against one point or other, and at last they hardly fired at all. But the case is different when the men are not allowed to fire independently; when each face of the square is divided into 4 platoons, every two of which act relatively to each other, and fire alternately when the ennemy is immediately in their front.

It is absolutely necessary in such cases that a few men should advance in front of each face of the square, about the centre, for the purpose of firing at the enemy's skirmishers and small parties; and that the men in the ranks should on no account be allowed to fire independently. When the square is put in motion, these skirmishers step forward, and two files are pointed out to them, who will open and allow them to enter the square when the enemy advances.

Hurry of movement is here attended with the disadvantages, pointed out in §. 147. to the greatest extent; and has frequently caused the *defeat of a square*.

When the square halts, 4 or 6 files on the right and left of each face will fall back, so as to cut off the angle, by which they will be less exposed to the enemy's artillery. If there are any guns, they should be placed in the angles of the square.

§. 151. *Conduct of Infantry when opposed to Infantry.*

a) *General Rules.* 1) Endeavour to fire before the enemy when near. 2) If you have advanced within the distance of 80 paces without firing---fire a volley and charge immediately. 3) If you have continued for some time firing, and approaching each other until you are within

40 paces---charge after firing. 4) Never charge with your pieces loaded, it induces the men to fire singly, instead of using the bayonet, and diminishes the impetus of the charge. 5) Never fire at infantry except by volleys of battalions, this produces double or treble the effect of the fire by platoons or files ; as the battalion will fire 5 volleys in a minute, while only two rounds will be fired by platoons. In addition to which, the fire by platoons and files has the disadvantages mentioned in §. 148. In firing volleys, the commanding officer is more master of his battalion, and the loading is completed before the word for firing is given. This kind of firing of itself prevents confusion.

b) *When attacked by infantry*, do not fire at a greater distance than 75 paces, unless the enemy commences his fire sooner, and the musquets being loaded with two balls, it is probable one volley will then break them. The Austrians acted in this manner against the Prussians at Breslau, after they had crossed the Lohe, and according to Warnery's account, two thirds of them were killed; but in this case it is necessary you should fire the very instant that the attacking party halts, or the enemy may be able to fire first. If the enemy, when attacking you, begins to fire early you must do so likewise; but even then it is better not to fire at a greater distance than 200 paces, by which you will do more execution.

c) *When attacking infantry*. 1) If the enemy is posted on open ground, and you are not provided with artillery---fire a volley at the distance of 300 paces, and then advance 50, and repeat it, &c. If you were not to fire at a long distance, you would hazard being defeated by the enemy's fire, when you approached nearer. 2) If both parties are provided with artillery, there is more reason for firing soon, and you should give time for your guns to fire. 3) If you are superior to the enemy in artillery, you should fire several times from the same spot, and only advance a short distance between each halt, for were you in the two last cases, to advance without firing, you would lose the advantage, which you may derive from your guns; and as the enemy's infantry would not have suffered, the fire of his musquetry upon your near approach, would become very destructive. The more you engage the enemy, while you are advancing, in the three last cases, the more you lessen the advantage he would otherwise have of being able to fire more frequently than yon. 4) If the enemy has artillery and you have not, or if his troops are covered---advance rapidly, and fire only now and then for the purpose of keeping the attention of your men occupied. Small detached parties and skirmishers should be advanced in front, who will fire at 400 paces distance to induce the enemy to fire in return, and will endeavour to make him continue his fire; unless this is done, you will most probably be routed by the more powerful fire of the enemy, when you approach within a short distance of him. 5) Never fire in any manner except by volleys of battalions. The firing by platoons will detain you longer on each spot, a great part of your men will at times be entirely idle, neither firing nor advancing; and an irregular fire will be the consequence, from which it will be very difficult to withdraw the troops and put them again in motion.

d) *When remaining Stationary while engaged.* If you receive the enemy's fire at the distance of 300 paces or less, and remain stationary while engaged, for the purpose of favoring another attack; the great object is to keep the men employed: the firing volleys is not so well adapted to this end: the 2 platoons of each grand division should fire alternately and slowly: in this maner a continued fire will be kept up in each grand division, and there will be little danger of falling into an irregular fire.

e) *Considerations respecting the Ground.* If you wait the enemy's attack, you should endeavour to gain a position where he may be retarded, at about 80 or 100 paces in front, by obstacles in the ground; for this purpose you should post yourself at the above distance behind ditches, swampy ground, low and thick bushes or hedges, (over which you have a distinct view,) rivulets, &c. and the moment you see the enemy impeded by these obstacles, fire a volley, and he will probably be thrown into confusion.

If you were to post yourself immediately behind these impediments, they would be totally useless. It is a considerable advantage in your firing, to be able to cover your men behind ditches, embankments, &c. or in pits, so that they have a clear view in front; but still the enemy must be met, when he approaches within 40 paces.

SECTION X.

Of Supplies.

CHAPTER I.

Of escorting Convoys.

§. 152.

GENERAL RULES.

a) *A knowledge of the Country* is obtained by maps; from persons in the army, who have been previously acquainted with it; and from country people procured for that purpose.

b) *The number of Waggons, the strength of the Escort, and the place to which the Convoy is to proceed,* must be exactly known. If possible, some empty waggons and spare horses should accompany the convoy, which will be of the greatest use in case of accidents, and in procuring forage. A supply of ropes, hatchets, lanthorns, nails, &c. is indispensible on the march.

c) You should learn, whether you are to carry forage with you, or to receive it on the march, and, if the latter, the places at which you are to receive it.

d) *Disposition of the Waggons.* The waggons of the greatest importance should be placed in the centre, intermixed with others of less consequence; it should however be concealed from the escort, which waggons contain money or articles of value.

e) *Distribution of the Escort.* One part remains with the waggons for the purpose of conducting and keeping them in order; another part reconnoitres the country through which the convoy is to pass; and the third and strongest part moves with the waggons for their protection and defence. The party which conducts the waggons, should be small, and should consist, if possible, of cavalry: trusty men should be chosen for this service, with as many non-commissioned officers, as can be spared; a non-commissioned officer with 3 or 5 men, should conduct from 10 to 20 waggons. If there are any of the waggon train, the party with the waggons may perhaps be dispensed with, particularly if the escort is weak.

The party, for the defence of the waggons, should be formed in 4 divisions; 2 of which are with the centre, 1 in front and 1 in the rear. *

Half, or two thirds of the parties in the centre are formed into small parties, which march on the flanks of the convoy, at from 200 to 400 paces from each other. The party which reconnoitres in front, should consist of cavalry; or, if the country is much intersected, of cavalry and infantry together.

When there are guns, they will be with the parties in front, rear and centre.

---

* By our regulations, the escort is divided into 4 parties, one of which forms the advanced, one the rear guard, and the 2 others are distributed along the convoy in small parties.

§. 153. *Examples of the Distribution of Escorts.*

1st. 200 *Waggons with an Escort of* 200 *Infantry.* Forty men will be attached to the waggons, divided into 10 parties; thus there will be 4 men to every 20 waggons. The remaining 160 men will be formed in 5 divisions; 2 of which will be with the centre, 1 in front, 1 in the rear, and the 5th will reconnoitre the country some 1000 paces in front of the convoy.

2d. 200 *Waggons with an Escort of* 200 *Infantry and* 100 *Cavalry.* 50 cavalry are with the waggons, formed in 10 divisions; 100 infantry are in the centre, 50 in front, and 50 in the rear; the remaining 50 cavalry reconnoitre the country, about 4 miles in front of the convoy.

3d. 1000 *Waggons with an Escort of* 2 *Battalions.* 100 men reconnoitre the country in front. 250 men are selected and divided into 50 parties, each party will have the conduct of 20 waggons, thus there will be 1 man to 4 waggons. Half a battalion is in front, and the same in the rear; and 1 battalion in the centre; this will allow of one platoon to every 2000 paces, to repulse the enemy's irregular attacks.

4th. 200 *Waggons with* 100 *Infantry.* 20 men reconnoitre the country in front, and the remainder are divided into 4 parties; each of which will have 50 waggons under its charge; the first party will march in front, and the last in the rear of their divisions of waggons. If the escort consisted of cavalry, the distribution would be the same as the infantry.

§. 154. *Conduct of the Men with the Waggons.*

Each man is to pay attention to the waggons under

his charge, and will be careful — *a*) that no intervals take place, — *b*) that the waggons form according to the order given — *c*) that the waggoners remain with their waggons, are not intoxicated, do not run away in case of alarm, and that both themselves and their horses are provided with food. If a waggon breaks down, it must be drawn aside, that the others may not be detained; 2 or 3 men only will remain with it to assist in repairing it; and if this is not possible, the load should be placed upon some of the other waggons.

§. 155. *Conduct of the Advanced Guard.*

a) *When the Advanced Guard consists of Cavalry.* It is formed in 3 divisions, 4 miles in front of the convoy: the 1st division is in the centre, and the others move in an oblique direction on the flanks; the whole observing the rules laid down for advanced guards in §. 26. The centre division causes the roads to be repaired, when necessary, by peasants or by pioneers, who accompany them. Entrenching tools are carried with the guard upon waggons for this purpose.

Different passages must be formed for the escort near small bridges, &c.

If forage is to be procured where the convoy halts, the advanced guard will make immediate arrangements for the delivery of dry forage, if it is to be had; or will otherwise look out for places where green forage can be obtained.

b) *If it consists of Infantry*, it will be formed, as above, 2 or 3000 paces in advance, and will perform the same duties as the cavalry. A few men should be sent on horseback into the villages, about 6000 paces distant in front and on the flanks.

c) *If the convoy is to proceed secretly,* there should be no parties on the flanks, nor at any great distance in front; and the precautions, laid down in §. 37. for secret marches, should be strictly observed.

§. 156. *Conduct of the Escort divided in Front, Rear and Centre of the Convoy.*

As soon as the convoy arrives in a part of the country, where it is likely to fall in with the enemy, parties must be detached into the villages, woods, &c. on the flanks.

On falling in with the enemy, the convoy must be immediately closed up; the cavalry will advance towards the enemy, while the infantry cover the formation of the waggons.

In some instances, particularly if the enemy is weak, the convoy should not be closed up; the enemy may probably endeavour to detain you with small parties, until he can bring up a stronger force. You may also be thus drawn too far from the army to which you belong. In such cases the cavalry must endeavour to keep the enemy at a distance, and the infantry must defend the waggons to the utmost of their power; but at all events, an immediate report of your situation should be sent to the nearest corps, garrison, &c.

§. 157. *Of the Formation of the Waggons when attacked.*

Each division causes the waggons under its charge to form in a line, a a. The waggons of the first division form at the distance of 20 paces from those of the second, b b. The third form in like manner.

The horses are placed between the lines of waggons a a and b b. &c. and the flanks of the intervals are each covered by 3 waggons, a b, a b, b c, b c.

When the convoy is large, a chain of waggons, d, d, d, d, should be formed at from 10 to 20 paces distance round the others, as above, the horses belonging to this chain should be placed with the others between the rows of waggons.

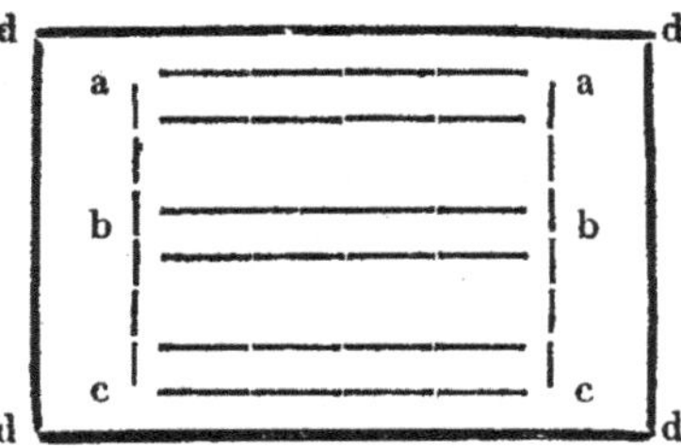

The infantry destined for the immediate defence of the waggons, is placed in the interval between the chain and the waggons in the centre. The waggons forming the chain stand lengthways, the shafts of one being brought under the body of the next waggon.

If the waggons, forming the chain, were placed with

the shafts either inwards or outwards, it would be easy for the enemy to break the chain, by drawing a few of the waggons out.

§. 158. *Of the Situation proper for forming the Waggons.*

The waggons should be formed where the flanks of one or two of the intervals may be covered from attack by impediments in the ground. Houses and walls, however, are not to be considered as impediments in this case.

§. 159. *Of the Defence of the Convoy.*

a) If the attack is made by cavalry, the infantry of the escort must be placed between the waggons, having the shafts in front of them. The artillery is stationed at the angles between the waggons.

b) The cavalry must remain without, and act separately; or is sent entirely away: but it always supports the infantry as far as lays in its power.

c) If attacked by infantry and artillery, the infantry of the escort must form round the waggons, being perfectly disengaged from them. If there are any intervals in the infantry, they must be closed up with waggons.

d) If the escort, consisting only of cavalry, is attacked by a superior force of cavalry, and has hope of being supported; the men must dismount and defend the waggons in the manner directed for infantry.

§. 160. *Conduct near Defiles and Woods.*

The country should be previously reconnoitred to

the right and left; and if there is any reason to apprehend that an enemy is near, the convoy should be formed in front and in rear of the defile or wood, which should be occupied by the troops; by this means the danger is lessened, as the convoy is close together and compact. Even if you form only half the convey, and then cause the leading waggons to pass the defile or wood, you have the convoy closer up, by half the extent occupied on the march, than it would otherwise be, and, of course, you are better enabled to defend it.

§. 161. *Conduct at Night.*

You form, if possible, on an island, a peninsula, or near a piece of water, &c.; post outguards, and observe the directions given for conduct when in quarters.

§. 162. *Conduct, with a Convoy of 1000 Waggons or more, in a Country where Flying Parties of the Enemy are to be apprehended.*

The convoy must be formed in 2 divisions, and the first division must commence its march. As soon as the last waggon of this division is in motion, it forms again, and the first waggon of the second division then sets out; the second division also reforms as soon as its leading waggon has reached the spot where the first division is formed, the leading waggon of which immediately moves forward again, &c. in this manner half the day is allowed for feeding the horses, and the column is never more than half the length it would otherwise be.

§. 163. *Of particular Cases.*

With convoys in rear of the army, small escorts only are necessary to preserve order, as they cannot push the advanced guards to any distance in front and on the flanks. If surprised by a party of the enemy, nothing is left but to abandon part of the convoy, and to prepare for the defence with the remainder.

CHAPTER II.

Attack of a Convoy of Supplies, escorted by a small Datachment of Cavalry.

§. 164. *March.*

You must approach by secret marches, and cause the convoy to be watched by persons in disguise; then prepare an ambuscade in that part of the country, through which the convoy must pass.

§. 165. *Attack with a small Detachment.*

The attack must be executed with the greatest rapidity, that the convoy may be thrown into confusion. The enemy must be beaten in detail, before he can form, or he will take advantage of his superior numbers.

When part of the convoy has passed a defile, it should be attacked, and the defile barricaded with waggons, in order that the parts of the escort, thus separated, may not be able to support each other.

Two sections attack the escort and one the waggons, which must be obliged immediately to drive off, or the horses should be hamstrung, or such waggons, as are loaded with gun-powder or forage, burned; the execution of the latter, however, requires some time. A quantity of straw should be placed under the waggons, and spread out to a little distance from them, and then set on fire; the men retiring immediately.

Enquiry should be made, which are the most valuable waggons; and the first prisoner taken must, by threats, be forced to point them out; these will of course be seized.

If the detachment is very weak, and cannot make an attack in a defile, without being discovered, still there may be a possibility, in some place or other, to get at the waggons, without being at first perceived, and to destroy part of them, to kill the horses, or, if they belong to the enemy, to cut the traces and carry them off; or to secure part of the waggons, before the escort can assemble to oppose it. A road, by which to retreat, must have been fixed upon beforehand, and the bridges put in such a state, that by removing a few planks, the pursuit of the enemy may be prevented.

§. 166. *Attack with a Detachment stronger than the Escort.*

In this c[illegible] a attack the escort only, and

endeavour to beat it in detail. The first attack should be made upon the centre, and when that is broken and the convoy separated, the business will soon be settled with the remainder.

If the convoy is defended by infantry, who form behind the waggons, you must cause your cavalry to dismount and attack on foot. This must also be done, if the enemy's cavalry dismount.

If you have guns with yon, fire upon the waggons, until you throw them into confusion.

SECTION XI.

Of seizing Patrols, Couriers, &c. and Levying Contributions.

§. 167. *Of seizing Patrols and Couriers.*

The march of the party to the places where the patrols and couriers may be expected to pass, should be conducted in the manner pointed out in §. 37 and 41.; there an ambuscade should be prepared (and, if necessary, at different spots) in such a manner as to command the road, by which the enemy's patrols and couriers pass, and to be able to cut them off, when they arrive at a certain point.

Part of the detachment must remain mounted within the ambuscade, while some men watch the road. No passengers or officers, &c. of the enemy, must be seized, or the detachment might be discovered, and the object not attained. The attack must not be made too soon, or the prey may escape; nor must the men fire; both these might cause a discovery. After the duty is performed, the detachment must not remain in the same place. You must make yourself previously acquainted with the places where the rivers may be passed, and also the spots in the woods and forests, by which the retreat can be made undiscovered.

Lieutenant-colonel Emmerick, who was frequently employed on these expeditions, during the seven years' war, marched in this manner, in November, 1761, with a small detachment of Hanoverian riflemen, from Einbeck, through the forest of Sollingen, towards Lipstadt, and on that side stole through the enemy's posts, (having succeeded in the same thing near Warburg) so as to get into his rear by way of Iserloh, Perleburg, &c. upon the road from Frankfort to Cassel. By means of persons sent out in disguise, he gained a knowledge of the roads, by which the enemy's couriers remount horses, &c. passed; he seized the former and dispersed the latter; and after executing any thing of this kind, he immediately removed to a station several miles distant, and remained concealed in some large wood. At length, he returned by way of Hœxter, on the 6th of January, without having lost a single man.

§. 163. *Of Levying Contributions.*

For this purpose, you must endeavour to seize some place which is defended by the inhabitants, or by armed peasants, or that is otherwise weakly garrisoned. You should approach it by secret marches and if you are too weak to undertake any thing else, attempt to get possession of one of the gates by surprise. At the same time, a part of the detachment should be formed in single rank, and should be moved about, so as to discover the men in different situations to the inhabitants; the drums should also beat, &c.; and you should procure carts which should be placed so as to be perceived by the inhabitants at a distance, in order to induce the belief that you have heavy artillery, &c.

If you are too weak even to adopt this method, you must regulate your march in such a manner as to arrive at the gate during the night; in front of which you must cause some of your men, together with some carts or waggons, to be seen moving backwards and forwards; if this does not effect your purpose, you must appear as if you were making preparations to use force, and cause some *marrons* to be exploded, to make them imagine you are going to destroy the gate with heavy artillery; at the same time, threatening to plunder and burn the town. The great object is, to conclude the business quickly and during the night; for which purpose, you must make the most of the garrison's first alarm, and having secured the contribution, retire as quickly as possible, without discovering your weakness.

If the contribution is to be raised in the country, the detachment should always remain concealed, (see §. 37, &c.) and frequently move from one spot to another, to prevent any body from being acquainted with its situation: a small party only should be employed to raise the contribution. Without this precaution, you might be seized by the enemy, who would be anxious to prevent you from effecting your design.

In every case you must cause the principal municipal officers to be brought before you immediately, and oblige them to deliver up some of the most respectable inhabitants of the place, as hostages.

Part only of the contribution should be required in the first instance, and the demand repeated, by which means you will be more likely to obtain your object.

PART THE SECOND.

INSTRUCTIONS FOR OFFICERS DETACHED WITH ARTILLERY.

SECTION I.

Conduct of an Officer detached with Artillery, in an open Country and of an Officer commanding a Detachment, with respect to Artillery.

CHAPTER I

List and Examination of Articles necessary for Artillery upon a March.*

§. 169.

a) *List of Articles belonging to the Gun and Carriage.*

2 Rammers with spunges.

* If every article is not inspected, according to a regular Roll, before commencing the march, something may be forgotten. How often has such neglect caused the failure of an expedition. In 1778, the artillery men, belonging to the detachment, which marched from Glatz, to support the block-house at Schwedelsdorf, forgot the case shot, the consequence was, that the whole detachment, together with the artillery, surrendered in a plain, to the enemy's cavalry.

2 Leather cartouches.
1 Box for quick matches or tubes.
1 Box for mealed powder (with howitzers only.)
1 Box for port-fires.
1 Ladle or wad-hook.
1 Worm.
2 Linstocks, with cocks.
2 Priming irons.
1 Gimblet.
4 Handspikes, (only 2 for a 3-pounder.)
1 Spade.
1 Shovel.
1 Pick-axe.
1 Hammer and 1 pair of tongs.
1 Axe.
1 Hatchet.
1 Drag chain.
1 Limber chain.
2 Pair of drag ropes, with pins.
1 Apron.
1 Tampion.
1 Level, or with a howitzer, a Quardrant.
1 Tar bucket.
1 Winch.
3 or 4 Rugs.
1 Truck.
1 Leather bucket.
1 or 2 Rough sheep skins, to repair the spunge.

b) *Ammunition.* There are usually 200 rounds with each piece of cannon, one third of which, at least, should consist of grape or case shot. For these 200 rounds, about 250 tubes, and from 10 to 20 pounds of matches are requisite.

A howitzer should have about 120 shells with their fusees, and about 40 rounds of case shot, besides about 6 carcasses, and the same number of light balls: these latter however depend on the object of the expedition. As many tubes, matches, and port-fires, are necessary as with guns.

If the howitzers are to be used as mortars, the following stores will be further required, for each howitzer.

1. Several empty shells (from 50 to 100 for each day's bombardment.)

2. The necessary quantity of powder for the above number of shells. With the 5-and-a-half inch howitzers, from 2 and-a-half to 3 pounds may be allowed for each charge; and with the 8-inch howitzers, 5 pounds.

3. Several pounds of tow, with scales and weights.

4. Fusees prepared, between 10 to 20 more than the number of shells.

5. Two mallets and fusee drivers.

6. One funnel for the shells, and with a 30-pounder, (or heavier metal) 1 chamber funnel, and 1 bomb chain.

7. Two or 3 pounds of mealed powder.

8. Light balls and carcasses, with wooden bottoms for the former. The number of carcasses will depend upon the object of the expedition. If it is intended to set fire to a magazine, &c. a greater number of these than of shells will be requisite. Rugs for packing, and sheepskins to cover the ammunition.

c) *Horses.* In bad roads 1 horse is necessary to every 200lb. weight of ordnance, therefore 6 horses will be required for a 6-pounder weighing 1200lbs. In some services, however, 1 horse only is allowed for 250 or 300lbs. 4 horses are then sufficient for a 6-pounder.

One horse is allowed for every 300lbs. weight of ammuuition: therefore the 200 ronnds for a 3-pounder require a waggon or cart with 3 horses; for a 6-pounder a waggon with 6 horses, and for a 12-pounder 2 waggons with 6 horses each.

One waggon with 6 horses is calculated for the transport of 20,000 musquet cartridges for infantry; to which must be added, 1 flint for every 40 or 50 cartridges.*

d) *Men.* If the guns are to be served quick, and the cannonade lasts some time, a 3-pounder requires 8 men, a 6-pounder, 12 men, and a 12-pounder, 16 men; but if the men are to draw the guns, 1 man must be allowed for every 60 lbs. weight. Thus the 6-pounders, weighing 1200 pounds, will require 20, and the 12-pounders, weighing 2400 lbs. 40 men.

There must be one driver to every 2 or 3 horses; one train non-commissioned officer to every 30 or 40 horses, and one train officer to every 100 horses.

In sieges, 4 gunners or bombardiers are necessaray to each piece of ordnance.

e) *Harness and Stable Implements for 2 3-pounders, 4 Carts, and an Ammunition Waggon.* Each gun has 4

---

* The general calculation on an average is for a carriage

With 4 horses	24 cwt.	
6 ditto	- 30 ditto	
8 ditto	- 36 ditto	
12 ditto	- 48 ditto	

If the roads are good, the horses strong, and the wheels in good order a carriage with 2 horses may convey 2000 lbs. with 4 horses 3600, and 6 horses 4800, for a short distance.

horses, each cart 3, and the waggon 4, making in the whole 24 horses and 7 carriages.

24 Bridles.
24 Halters, with chains.
24 Sets of thill harness.
10 Sets of trace harness.
32 Pair of traces.
7 Sets men harness.
4 Single bars (1 for each cart.)
3 Double bars (1 for each gun and 1 for the waggon.)
2 Setters.
8 Saddles (1 for each driver.)
8 Whips (1 for each driver.)
8 Curry-combs and brushes (1 for each driver.)
8 Sieves.
24 Corn sacks.
8 Water buckets.
8 Forage cords.
7 Pair of long reins (1 to each carriage.)
24 Nose bags.
8 Mangers (1 for 3 horses.)
30 Picket poles (1 for each horse, and 1 besides to every 3 or 4 fathoms picket rope.)
12 Fathoms of picket rope (4 feet to each horse.)
1 Chaff-cutter, with 2 knives, a whetstone, &c.
2 Corn measures.
4 Sickles.
24 Horse-shoes, with nails.
7 Lanthorns.
7 Grease boxes, (1 for each carriage), and when the expedition is to a distance, a quantity of grease.

1 Hatchet and a pick-axe for each carriage, (the hatchets, necessary with the guns, have been already mentioned in a.)

In addition to the above, a sufficient stock of straps, cords, ropes, nails, saws and iron clouts, should be provided.

f) *Camp Equipage. &c.* Tents, with tent pins, for the non-commissioned officers, gunners and drivers, in the proportion of 1 to 5 men; with each tent, 1 canteen; 2 blankets; 1 camp kettle and frying-pan; 1 bag with straps for the tent pins, and 1 hatchet, if there is none with the carriage; and 1 haversack to each driver.

g) *Forage Waggon, Bread Waggon, and Bat Horses*

1) It is unnecessary to carry forage, except when it is not to be procured in the country, in which the party is acting. A sufficiency for 3 or 5 days may be conveyed on the gun-carriages and ammunition waggon: the allowance is, for each horse from 8 to 10lbs of corn, and 10lbs. of hay per day; making together about 20lbs. One forage waggon with 4 horses carries 1500lbs. (75 rations) If, therefore, a forage waggon is allowed to each 3-pounder, forage will be provided for 7 days (that is reckoning 4 horses for the gun, 3 for the ammunition cart, and 4 for the forage waggon, 11 rations will be required per day, (which divided by 75 gives 7) and forage for 3 days being transported on the gun carriage; there is thus a supply on the carriage of the gun and the forage waggon for 10 days: besides which, if rough forage, as is generally the case, can be procured, there is provision for 20 days, as corn need then only be carried, and of that of course a

double quantity. If there is only one forage waggon, a supply can be carried with a 6-pounder, but for half, and with a 12-pounder, but for a quarter of the time.

In summer and autumn, when supplies are every where to be found, forage waggons are, of course, not required.

2. With respect to bread, 2lbs. per diem are necessary for each man. One waggon with 4 horses carries 750 rations, the supply of 150 men for 5 days, or 75 men for 10 days. A bread waggon is always necessary with 150 men, as the bread must be sent for to the bakery, which is often 2 days journey distant. If the expedition is to last 10 days, one bread waggon will be sufficient with 150 men; every man being provided with bread for 5 days, and the supply for the other 5 days being carried on the waggon. If the expedition is to last 15 days, an additional bread waggon will be requisite.

3. Bat horses are used for the conveyance of the tents. One horse for 6 or 8 tents.

§. 170. *Articles required with a small Park of Artillery.*

When either 1 or more brigades are detached from the army for any space of time, they must be provided with all the articles necessary for the use and service of artillery.

When the division detached consists of 20 pieces of artillery, the following articles will be required, in addition to what have been mentioned in the foregoing paragraph.

1. Two spare carriages.

2. Four spare wheels, and as many spare axletrees, and one pole; to be conveyed on the spare carriage.

3. One forge cart with 2 waggons, containing coals and tools for the use of the forge, sufficient for 1 master and 2 assistants, together with a stock of nails and iron; and also the implements necessary for making shot red hot, as grates, tongs, &c.

4. Two waggons, with medicine for horses, leather and tools sufficient for 2 or 3 sadlers.

5. One or 2 waggons for the wheelwright.

6. One waggon with a gin and stock of rope.

7. Two or 3 waggons of cart grease.

8. Two or 3 waggons, containing the tools and materials, required in making and repairing cartridges, carcasses, fire-balls and fusees. The number of these articles cannot be determined, as it must depend on circumstances. If the division is to act against a fortress, and is accompanied by howitzers, and if there is not a sufficient stock of carcasses, materials must be conveyed for composing them, that those which are expended may be replaced.

9. From 10 to 20 waggons with the reserve ammunition, and between 1000 and 2000lbs of powder for fougasses, &c. The ammunition must be accompanied by a sufficient number of port-fires, tubes and matches, according to §. 169.

10. Two or 3 waggons with entrenching tools, spades, axes, hatchets, bill-hooks, plummets, masons levels, tracing lines and measuring rods.

The articles necessary with each carriage, both for the field and the stable, are pointed out in the foregoing paragraph.

§. 171. *Examination of the Articles to accompany the Detachment.*

a) It is necessary carefully to examine: 1st. That the gun is not damaged. 2d. That the vent is not enlarged by use. 3d. That the elevating screw is in good order; and 4th. That the spunge fits the bore.

b) Further, it should be ascertained, that the tubes are not too large for the vent, and that the balls and cartridges fit the bore, and allow the proper windage. To those, who are not aware that 12-pounder cartridges are not unfrequently issued to a detachment of 6-pounders, and that the ammunition supplied is often unserviceable, this examination will appear superfluous.

§. 172. *Of particular Circumstances.*

If the detachments are to traverse marshy ground, or hills, over which are no regular roads, application must be made to the park for ropes, blocks, tackles, handspikes, windlasses, and an additional supply of spades, axes and hatchets, in order to remove such impediments and difficulties as may occur.

CHAPTER II.

Of the March.

§. 173. *Conduct and Order of the March.*

a) The artillery is in front of the columns, always however preceded by an escort.

b) If there is a supply of ammunition on the limber, the waggons will follow the column; otherwise 1 waggon at least must accompany every 2 or 4 guns.

c) The artillery-men will be with the guns, except a few, who must remain with the ammunition waggons. The commanding officer is in front of the leading gun, with some non-commissioned officers, in order to discover and point out the best road.

d) When near the enemy, the cartouches must be filled (especially with 6 and 12-pounders, where no ammunition is carried with the guns) and placed on the gun carriages, the matches must be lighted, and every thing prepared to unlimber.

e) If the enemy annoys the flank of the column, 2 guns will be brought to that side, and advanced with drag-ropes. They will fire with caseshot only, and never at a greater distance than 400 paces. The second gun will not fire until the first is reloaded.

§. 174. *Rate of March.*

If it is required to fix the time of arriving at any place, it is necessary to observe, that, when the horses can start fresh, and rest immediately after finishing the work, they may, if the roads are not too bad, perform 2 miles (the mile equal 2500 paces) in half an hour.

4 miles in 1 hour and an half.
8 ditto 4 hours.
16 ditto 10 ditto.

In the camp near Hannover, in 1791, the horse artillery performed 350 paces in a minute, 3500 in 9 minutes and a half, 6000 in 22 minutes, and 12,000 in an hour.

§. 175. *Respecting a Gun being overturned, set fast in the Road, or sunk in a Swamp.*

a) If a 3 or 6-pounder is overturned, the gun should be fastened with a rope or chain to the carriage, and then righted.

If a 12-pounder is completely capsized, the gun should be taken off the carriage, which should be set upright; both the wheels are then taken off, the gun replaced on the carriage, and the wheels are put on again one after the other, the carriage being raised with a windlass or handspikes.

b) *If a wheel is set fast in a rut of stiff earth or clay.* A rope is tied round the top of the felloe, behind the spoke, the knot, (which should be on the outside) must be such as to loose of itself when the wheel turns; a horse is then attached to this rope and draws with the others.

If the men do not know how to tie a knot of this kind, a hook should be substituted, which must be held in its place while the horse is drawing.

c) *If a wheel gets jammed between frozen ground, or in a rock,* the impediment must be removed with a pick-axe; one of which should be with every gun.

d) *If a gun is sunk in a swamp.* The jack should be placed on a plank or some handspikes, &c. that it may not sink, and applied under the axletree of the gun, which should be raised at the same time that the horses are made to draw. If the jack cannot be used in this manner, and the gun is sunk very deep, it will be necessary to dig round the wheels, and slope the ground away in front, upon which fascines or planks should be laid.

e) *If any accident whatever occurs.* The men and horses of several guns should be brought to assist.

§. 176. *To get a Gun up a very steep Hill.*

1. If the top of a hill, up which the horses cannot draw the gun in the usual manner, is an extensive flat; the limber, with the horses, must first be got up, to which the axletree of the gun must then be made fast with a rope, (the muzzle of the gun being turned towards the limber) and 2 or 3 men placed with handspikes at the trail; the remainder will be placed at the wheels to push them forwards when the horses draw.

Handspikes must be laid on the ground wherever the rope trails.

2. If the summit of the hill is not flat, and the limber can move only to one side, a strong pole must be fixed 2 or 3 feet deep in the earth, on the top of the hill, to which a block must be attached, and the rope passed through it.

a b is the summit of the hill; at *a* is the pole with the block; *c* is the gun from which the rope passes through the block *a* to the limber, with horses, at *e*, which moves towards *b*. If there are trees on the top of the hill, the block may be fixed to one of them.

3. If the hill is of such a nature that the limber with the horses cannot be got up it, the pole with the block should be fixed at the top, and a rope from the gun passed through the block, and then carried down the hill to the limber, which will move off from the hill, and, of course, the gun will be drawn up. It is to be understood that the men are to be employed at the gun in assisting to get it up the hill.

If *a* is the summit of the hill, where the pole with the block is fixed; *c* the gun, and *d* the limber and horses; then the rope will be carried from *c* through *a* to the limber at *d*, which will move towards *e*.

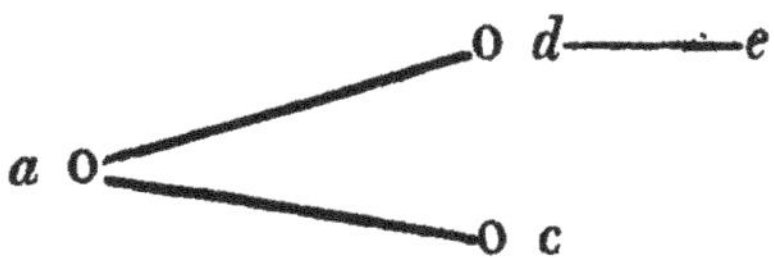

These expedients are only necessary where the ground is extremely steep. If the gun is to remain on the side of the hill, a beam of wood should be run through the wheels to prevent their turning, and a post driven in behind the trail: as it is of the greatest importance to secure the gun from running back, if the means above mentioned are not sufficient, a rope or chain should be passed round the gun and fastened to a post above.

§. 177. *To transport a Gun across a River, hollow Way, &c.*

a) Some place should be sought, by people in a boat, where the bed of the river is tolerably even: here the banks should be sloped off on each side, and the horses got over; they may either be swum over, or, if the river is not very large, a place may probably be found, where this will not be necessary, which will generally be in an oblique direction. When the horses are over, a rope should be fastened to the arms of the limber, (the thill having been first taken off) which is then drawn across. A rope is passed round the axle-tree of the carriage, and another round the gun, both of which are made fast on the bank of the river; that there may be less difficulty in recovering the gun if it should be upset.

b) The quickest and the best manner of getting a gun over a hollow way, ditches, or small rivulets, is by sloping away the banks on each side, and throwing in the earth that is removed. If the ditch, &c. is very deep, and the banks cannot be well sloped, it should be filled with fascines, &c.

In these cases, strong poles and planks will frequently be of great service (when near a village) in forming a bridge, for the support of which a waggon may be used.

§. 178. *To get a Gun on the Top of a Tower or House, &c.*

a) This may be done by passing a beam through the wall, near the top, and fastening the longest end within the house; a block must then be fixed at the shorter end; and a rope passed through the block, one end of which will be fastened round the gun, and the men will hawl upon the other. For this purpose, the weight of the men must be equal to that of the gun.

b) Or it may be effected with two side beams and a roller; the ends of the beams with the roller being allowed to project over the wall, and being fastened within by ropes; one end of a rope is tied round the gun, the rope is passed through a block, and from thence round the roller, which being turned with handspikes, like a windlass, the gun is drawn up.

If a machine of this nature cannot be procured, a piece of wood may be used as roller, having a stay placed behind it.

CHAPTER III.

Of Artillery in an open Country.

§. 179. *Of placing the Guns.*

With Respect to the Ground. When you wait the enemy's attack in a position, it is necessary to consider,---

a) That hard, pasture and meadow ground, are favourable to the effect of artillery.

b) That the most advantageous situation is an elevated spot, sloping gently down towards the enemy.

c) That, if the ground in front of a gun is uneven with small hillocks, pits, &c. very little effect will be produced by caseshot at a short distance.

d) That guns should be placed at the distance of from 300 to 600 paces behind hedges, ditches, defiles, &c. so as to be out of the reach of the enemy's musquetry, but that he may be under the fire of your caseshot, while he is passing the obstacle.

e) That small hillocks, ditches, &c. should be made use of to cover the guns. Part of the earth, where the gun is placed, may be removed and thrown up towards the enemy, so as to form an epaulement, over which the gun may be fired: this should be particularly attended to in small hills and hillocks.

f) That, if it is purposed to defend a defile, a position should be taken up about 300 paces in rear of it; but not at so great a distance when the defile is some hundred paces in length, when the flanks of the guns are not sufficiently covered, or when there are only one or two guns, and not many troops for their protection; as under these circumstances, the enemy might pass the defile rapidly, and get upon your flanks. The defile should be rendered, in all cases, as difficult as possible.

g) That, when ordered to prevent the enemy passing a village, you should, if its situation is not too commanding, place yourself behind it, in order to throw him into confusion by a fire of caseshot, before he can form.

h) That, if you are obliged to wait an attack upon a hill, the guns must be so stationed, as to command the foot of it. The fire of grape only will decide the business. If situations on the top cannot be found for the guns, from whence the foot of the hill may be seen, they must be moved lower down.

§. 180. *Disposition of the Artillery.*

a) When detachments, on a small front, act on the defensive, having their flanks appuyed, the guns should be upon the flanks.

b) In acting offensively, 2 or 3 small batteries should be made to direct their fire upon the point of attack. The most decisive effect may be expected from a sudden, brisk and concentrated fire.

c) When the front is extensive, the artillery should be distributed along the whole in batteries, at the distance of 800, or at the utmost, 1000 paces from each other; not less than 6 guns should be placed in each battery. A scattered fire is not decisive, does not break the enemy's ranks, or compel him to retreat. The regimental guns should be on the flanks of the battalions.

d) In acting on the defensive, the artillery should be so disposed, that each battalion may have an equal number of guns with it, particularly when expecting the attack of cavalry. The artillery break the shock by the fire of caseshot, and infantry being near at hand, their fire may then complete the overthrow.

e) The guns of the largest calibre should be placed at the weakest points, that is, on the flanks, &c. provided they are not obliged to move frequently.

f) If any reserve is formed, it must consist of light guns, and must be kept masked, until called into action, near the weakest points, or where success is expected.

§. 181. *Observations on Forming.*

a) *If possible, the artillery should not discover itself*

until within 1200 *paces of the enemy*, till then the guns should be masked by troops, or concealed behind hills, &c. its sudden effect then produces fear and panic; while a protracted cannonade, doing little execution, encourages the enemy, and teaches him to despise the fire of our artillery.

b) *When circumstances will permit, the guns should be unlimbered at* 2500 *paces distance*, and then drawn forward by horses. A 3 or 6-pounder will require two, and a 12-pounder four horses, which are brought, with the swing bars, in front of the muzzle of the gun; a chain is slung round the axletree, the end of which is fastened to the bars; the trail of the carriage must at the same time be directed by some men with hand-spikes.

If a gun is unlimbered, while the enemy's artillery is playing upon it, it should not be turned towards them before it has been unlimbered; because, if a gun is turned with the horses, confusion is likely to ensue, if one of them is killed, &c.

c) *The intervals between the guns should be at least* 10, and if there is sufficient room, 20 paces or more; in the latter case, they will suffer only half as much from the enemy's fire as in the former.

If the guns are not placed at intervals of 20 or 30 paces, they will be exposed to the danger of being dismounted, and the men will be sacrificed unnecessarily. Hardly any circumstances can exist, in which it would not be adviseable to place the guns with such intervals; the apprehension of cavalry being able to penetrate between them, will only be entertained by those who are unacquainted with the effect of artillery.

§. 182. *Of serving Guns in Action.*

Three main points are to be attended to, viz.

No. 0. Pointing.
No. I. Spunging and ramming.
No. II. Handling the cartridge.

The gun is pointed by the serjeant only, who has No. 0, or rather no No. If he is also obliged to put in the tubes and fire, he will require at least 2 men to keep the matches or portfires in order for him, and hand them to him: one or 2 men are stationed near the trail of the carriage, with handspikes, to traverse the gun (upon a signal from the serjeant) either to the right or left. Thus the serjeant has 4 assistants.

Should the serjeant fall, he must be replaced by one of the men at the trail, who should therefore be competent to the performance of this duty.

A man ought to be added to No. I, either to relieve or replace him. The duty is performed by both together.

No. II. has 2 assistants to fetch the cartridges, and to relieve or replace him; they are responsible for the handing of the cartridges.

Thus a field-piece cannot be served without 4 gunners, and a few assistants, who are, in some degree, instructed in the duty of a gunner.

The serjeant puts in the tube, after the cartridge is rammed home and pressed to the bottom of the gun; so that the tube may not be put behind the cartridge: but to prevent the tube taking fire by a spark from the linstock, the serjeant puts his left hand upon it, and receives the match with his right. As soon as No. I. has rammed home, he gives the word, Fire! but the serjeant should always keep his eye on No. I. as he may not hear the word.

When tubes are used, they should be put in at the moment when No. I. draws out the rammer, who is not then exposed to any danger. The serjeant must be very careful, particularly after several shots have been fired, that he does not fire until No. I. has withdrawn the rammer, without the greatest caution, hurry is too liable to take place in this instance.

No. I. must keep his eye constantly on the mouth of the gun, that he may not step forward to spunge until it has gone off; as from the noise of the other guns he cannot always rely on distinguishing the report. The serjeant, at the moment he fires, must step aside, sufficiently to avoid the gun's recoil.

All these particulars must be strictly attended to, or the most fatal accidents are unavoidable.

It would also be dangerous to advance with the guns loaded.

When a gun has missed fire twice, the balls are drawn with the ladle, and the blank cartridge with the wad-hook, after the quickmatch or tube is taken out of the vent. But No. I. must not attempt to do this without being ordered. If the serjeant thinks that the tube has not entered the cartridge, he gives the word, *Ram home!* (after the tube has been taken out of the vent) he then puts in the tube, when the cartridge is rammed home.

The consequences may be fatal, if No. I. steps forward before he is ordered*.

* The above rules for precaution are equally applicable, when the serjeant does not fire, as is the custom in some services.

§. 183. *Advance and Retreat.*

When advancing near the enemy, the gun is drawn forward by men. See §. 167.

The gun is drawn forward by drag-ropes fastened to the axletree and carriage. Great assistance may likewise be gained by slinging a rope round the centre of the axletree, and fastening booms, of about 8 feet long, at 6 feet distance from each other; 4 men being employed at each boom.

aa is the axletree, *bb* the drag rope, *cc* and *dd* are the booms.

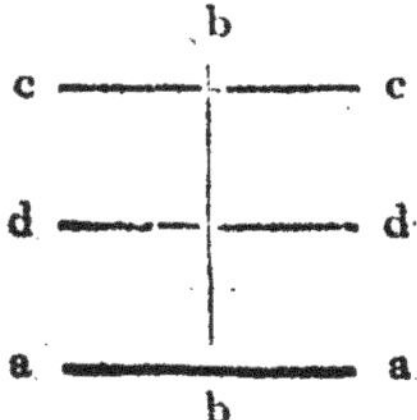

On halting, the men throw down the booms, and fall back immediately to the sides of the gun.

Thus from 16 to 24 men, besides the usual number, may be conveniently placed at a drag-rope, 20 feet long, without any confusion taking place.

For advancing with horses, see §. 181, *b*.

That the advance may be conducted with order, some men must remain with the horses.

In retreating, a drag-rope is fastened to the trail and hung over the pintle; when the limber is put in motion, the trail will drag along the ground, and 2 men will keep near it with handspikes, to be ready to raise it, if it meets any impediment on the ground.

This is a convenient method of retreating, as on the halt, the guns can immediately fire, the muzzles being towards the enemy.

§. 184. *Respecting the Distance of the Enemy. Range of Shot and its Effect.*

a) *Extreme range, but the effect of which is inconsiderable.* A ball may be thrown to the distance of 4000 paces with a 12-pounder; of 3500 paces with a 6 pounder; 3000 with a 3-pounder, and 2000 with a 5 1/2 inch howitzer, with a charge of 2 lb.* if the breech of the gun is depressed as much as possible, and the trail sunk a little into the ground, that is, if the gun is elevated 15 degrees. But the ball then moves in a very large curve, and does not ricochet. Some of the balls will fall several hundred paces nearer, and some as much farther; and perhaps a whole battery might fire for several hours at an enemy's camp, at the above-mentioned distances, with-

---

* See Manuel for Officers, vol. 1st. page 166, only guns of 18 calibers in length, with a charge of 1/2 the weight of the ball, and guns of 22 calibers in length, with a charge of 1/3 the weight of the ball, will throw shot to the above distances; a 3-pounder, 16 calibers in length, with a charge 1/3 the weight of the ball and elevated 15 degrees, will throw a shot only 2100 paces, etc.

When the experiments were made at Douay in 1771, the French 4-pounders, with a charge of 1/2 the weight of the ball and elevated 15 degrees, carried 3400 paces.

The range of a 42-pounder, with a charge 1/3 the weight of the ball and elevated 15 degrees, is 4400 paces, with the same charge and 43 degrees elevation, it is 5400 paces; and with the charge 2/3 the weight of the ball and 45 degrees elevation, it is 5600 paces.

With a charge 2/3 the weight of the ball, an 8 pounder, elevated 45 degrees, will carry 4100 paces, and a 4-pounder elevated 45 degrees, 3800 paces. Even musquets elevated 35 degrees will throw a ball 1300 paces.

The greatest range of a 5 1/2 howitzer, with a charge of 2lbs. is, when elevated 20 degress, 2400 paces, and above 3000 at 45 degrees elevation.

out a single shot taking effect. Still, however, an enemy may not consider this, and may alter his position, when he finds that he is within reach of our artillery. Examples of this do exist.

b) *Distance at which good effect may be obtained with round shot.* If the enemy attacks you upon a plain, in rough heathy, cultivated or light undulating ground; in any of these cases, the 3-pounders may commence firing with effect at the distance of 1700 paces, the 6-pounders at 2000, the 12-pounders at 2500, and the 5 1/2 inch howitzers at 1600 paces, the guns being laid at the line of metal elevation, and the howitzers being elevated 2 or 3 degrees, and having a charge of 2 lbs; the first graze of the shot and shells will be at the distance of from 700 to 900 paces, and they will ricochet to the above distance *.

The same extreme range will be obtained, either by directing the gun so as to give the 1st graze at 100 paces distance; that is, when it is laid at 0 degree, or

* When the experiments were made at Vahrenwalde, near Hanover in 1785, 24 rounds were fired from a 3-pounder upon rough heathy ground, out of which 14 reached the above distance and even farther, one fell short and fixed in the ground at 1300 paces, and some reached 1920.

Out of 18 shot fired from a 6-pounder, 11 ranged nearly 2000 paces, and some even 2100, but others fixed themselves in the ground at 1500 paces.

Out of 16 shot fired from a 12-pounder, nine ranged about 2500 paces, some nearly 3000, and others not more than 2000 paces.

In each case, the charge was 1/2 the weight of the ball, and the length of the gun 18 calibers. The range will not be so great with a shorter gun and a smaller charge; 3-pounders 16 calibers in length and with a charge 1/3 the weight of the ball, will throw a shot only 1100 paces; but with an elevation of 4 or 5 degrees, it will range 1800 paces, but only on favorable ground and when it ricochets high.

A 24-pounder at the line of metal elevation, ricochets generally 3000 paces, with a charge of 1/2 the weight of the ball.

horizontally; or by elevating the gun 2 or 3 degrees above the line of metal elevation.

If the enemy approaches nearer than above mentioned, that is, within 1200 paces of a 3-pounder, 1500 of a 6-pounder, or 1800 of a 12-pounder, the guns must be directed upon the ground at the distance of 100 paces; or the dispart must be added to the muzzle, and the gun then pointed direct upon the enemy.

If the enemy approaches within 1000 paces, the howitzers must be laid at the line of metal elevation, and cartridges containing of 1 1/2 pound of powder be used; the shell then makes its first graze at from 300 to 400 paces, and continues to bound towards the enemy; the first bound being very high, and the others less.

Though several of the shot and shells will not take effect, yet, on an average, about 1/3 will, which is very good effect at such a distance.

c) *Distance at which large grape shot may be used with considerable effect against an enemy who is stationary.* The guns must be kept at the above elevation until the enemy approaches within 750 paces of a 3-pounder, or of a 5 1/2 inch howitzer; 850 of a 6-pounder, or 1000 of a 12-pounder:* large grape should then be used, provided the ground is not too steep and irregular. **

---

* At these distances a few of the balls will hit a squadron. According to the experiments made in France, 7 or 8 balls will take place out of a grape consisting of 41 balls. But in the Hanoverian service, as the grape consists of only 30 balls, fewer will take place, but each ball being larger, will have greater effect.

** It is supposed that the charge is 1/2 the weight of the ball, and that each shot in the grape, weighs 3 1/2 ounces, with 3-pounders; 7 1/2 ounces with 6-pounders; 15 ounces with 12-pounders; and 6 ounces with a 5 1/2 inch howitzer. If the shot weighs only 3 1/2 or 4 ounces each, a 12-pounder will not

But as it is necessary to be sparing of the grape, that sufficient quantity may remain against the enemy's nearer approach, it is generally right to use it only at shorter distances. Thus an officer, who is detached, particularly in an open country, will do well to exchange with the reserve artillery so many round shot for grape, as will give him at least an equal number of each. Grape only will be decisive in the open field, round shot seldom take effect.

d) *Distance at which small grape may be used with effect, even against an enemy when moving.* When an enemy approaches within 500 or 600 paces of a 6 or a 12-pounder, small grape may be used, of which each shot, in the Hannoverian artillery does not weigh more than 3 1/2 ounces, or sometimes 2 ounces. Many of the shot will lodge in the ground, but from the number, considerable effect will be obtained.*

---

throw them more than 750 paces. If guns have a charge 1/2 the weight of the ball, and are less than 18 calibers in length, or if the charge is 1/3 the ball weight, and the guns less than 22 calibers in length: for instance, if a gun 16 calibers long, has a charge of only 1/3 the ball weight or less, even the last mentioned distance will not be reached with grape.

Owing to the uncertainty of round shot, it is often necessary to begin firing with grape very early.

To fire with round shot upon an enemy at the distance just mentioned, the gun must be pointed with the greatest correctness; if it is pointed directly at the enemy, the shot will make the first graze at from 700 to 900 paces, and then bound between 500 and 600 paces, and so high as to pass over either a battalion or squadron. If it is required that the shot should strike the enemy at the first graze, the gun must be pointed with the greatest nicety and the distance exactly known, or the first graze must be observed and the necessary correction made accordingly, but it is impossible to attend to all this in action, either when attacked or attacking, the smoke and a hundred other things prevent it. The main object for artillery in the field is to have large grape shot and a sufficient supply of it.

* At 600 paces (the pace being equal to 2 2/3 Hannoverian feet) 5 shot out

In the defence of a defile, or from an ambuscade, &c. when it may be necessary to fire at a less distance than 300 paces, the guns may be loaded with 2 small grapes, the balls weighing 3 1/2 ounces each, (those weighing 2 ounces would not have sufficient force) a double effect may then be obtained, and the carriage very seldom injured.

The guns and howitzers should be laid point-blank, until the enemy approaches within 300 paces; the guns should then be directed upon the ground at about 100 paces distant, or should be laid parallel to the surface of the ground on which they stand.*

c) *On commencing firing with grape, under particu-*

---

of a grape as above, viz. each shot weighing 3 1/2 ounces were put into a target 9 feet high with a 3-pounder, many of the shot did not take place, as the target was not more than half the length of a squadron, and the ground was the most uneven that could be found. With a 6 pounder, the grape containing twice the number of balls (60) 10 shot would certainly have taken place, and with a 12-pounder, the grape containing four times the number of balls (120) 20 would have struck the target, and a still greater number had it been the full size of a squadron.

At 400 paces, 10 balls with a 3-pounder, 20 with a 6 pounder, and from 30 to 40 with a 12 pounder struck th e target.

In the Danish artillery the grape for the 3, 6 and 12-pounders contain 100 balls, with which at 700 paces, the 12-pounder will place 14 balls (of 4 ounces each) and the 6-pounder 16 balls (of 2 ounces each) in a target 9 feet high on level ground. At 400 paces the effect will be about double. A 3-pounder will put from 21 to 24 balls (of 1 ounce each) into the target at 500 paces.

* When our experiments were made in 1791. Out of 6 rounds fired with grape, the guns being elevated 2, 3 or 4 degrees, 13 or 14 balls struck the target 600 paces distant. But when the guns were laid at the line of metal elevation 29 and sometimes 31 balls took place out of 6 rounds. At 400 and 500 paces, the guns had the greatest effect when laid point-blank. At 300 paces, the guns being laid at 0 degree, 68 balls took place out of 6 rounds; and at 1 degree only. 58. In the French artillery the effect was found to be the same; but in several other services they make use of elevations as shewn in the appendix.

lar circumstances. If well provided with grape, you may, on being attacked by cavalry, begin to use it, at the distance of from 700 to 1000 paces, or even farther. But if not more than 20 rounds are left for each gun, it should be reserved until the enemy's cavalry approaches within 500 or 600 paces.

When attacking, or when attacked by infantry in an open country, it is always to be expected that both sides will become stationary and continue the action, when within 500 paces of each other; in this case the grape should be reserved, unless you suffer too much by the fire of the enemy's grape before: by reserving the grape, the advantage is obtained of striking a sudden blow, which will create fear and panic, §. 181. But if the enemy's infantry is covered by bushes or otherwise, and that you cannot approach within 600 paces on account of his fire, it then becomes necessary to employ grape at the distance of from 800 to 1000 paces. This was done by the Hannoverian artillery at Bergen, against the French who were posted in the gardens near that place and near Crefeld, on the right flank of the allied army; and a very decisive effect was in a short time produced. Round shot will not obtain the same advantage in so short a time.

If attacked by cavalry in retreating, grape should not be used until they approach within 500 paces, its effect will then be very considerable, and the enemy will be fearful of again advancing so near.

§. 185. *Of the Distance at which to commence firing in a hilly Country.*

In the preceding paragraph the ground was

supposed to be sufficiently level for the balls to ricochet; but when the enemy is posted on a high mountainous ridge, &c. which will not admit of that, you must not expect to obtain a range of 1700 paces with a 3-pounder, of 2000 with a 6-pounder, or of 2500 with a 12-pounder; an important effect cannot be expected at a greater distance than 1200 paces.

At that distance, it is possible to perceive the graze of the ball, and the gun may be laid with correctness accordingly: an elevation of 1 or 2 inches should then be given to the gun, and if, after firing 5 or 6 rounds, the balls are observed always to graze in front of the enemy, the elevation must be increased; and if the balls then do not graze in front of the enemy, but go over him, the elevation must be diminished: lastly, when 1/3 or 1/2 of the shot strike in front of the enemy, the elevation is correct, and must be preserved until the enemy approaches nearer. When the enemy is above 1200 paces distant, the graze cannot be seen, and the true elevation cannot be found.

In this latter case, the 5 1/2 howitzers are most useful: it may be seen where the shells fall, and an effect will be produced by their explosion. At 10 degrees elevation, the 5 1/2 inch howitzers, with a charge of 1 1/2 pound, will throw a shell 1500 paces, and, with a charge of 2 pounds, 1800 paces. The elevation may be encreased in proportion to the distance, but never diminished, as the shells will then ricochet more.

Even when the gun is most correctly laid, not more than every 4th or 6th shot will strike a battalion; this is the utmost that can be expected.

As the enemy approaches nearer, the elevation of the

guns must be diminished; at 750 paces the 3-pounders, at 850 the 6-pounders, and at 950 the 12-pounders must be laid point-blank without any elevation; half the number of shot, particularly from a 3-pounder, will strike a battalion. *

On ground of this nature, the grape should not be used until the enemy is 2 or 300 paces nearer, than on a plain, and at least 1 degree more elevation should be given. At 700 paces, 1 1/2 inch elevation must be given to the guns and howitzers, and at 400 paces they must be laid at the line of metal elevation.

In all these cases, the effect of the guns is much less than on a plain.

---

* The following table will more particularly shew the ranges and effect of our guns:

Elevation.	Range.			No. of balls that will take effect against	
Degrees.	3-pounder.	6-pounder.	12-pounder.	Infantry.	Cavalry.
1	750	850	950	1/2	3/4
2	1050	1200	1350	1/4	3/8
3	1350	1550	1750	1/6	1/4

At 1 degree of elevation, a 3-pounder will throw a shot 750 paces, and 1/2 of the number of balls fired will strike a line of infantry, and 3/4 will strike a line of cavalry. Therefore, if 12 shots are fired, 6 will take effect against infantry, and 9 against cavalry.

But it is supposed here, that the distance of the enemy is known, or that the graze of the ball can be seen, which is seldom possible at 1200 paces, even when the weather is clear and the ground dry; therefore at 1550 or 1750 paces this effect will not be produced, unless the ground has been measured.

§. 186. *General Observations respecting Artillery in Action.*

a) The batteries are placed at from 30 to 80 paces in front of the line, so as to be clear of the points of alignement, and be enabled to fire to the right and left.

b) You endeavour in the first instance, if circumstances will admit of it, to ascertain the distance of the surrounding objects, as bushes, &c. which will be of great assistance in judging the distance of the enemy.

c) The fire should not be opened before every thing is in perfect readiness, so that when it commences, its effects may be more important, and also that the enemy's fire may not be attracted before it can be returned.

d) In the open field, each gun should be fired according to the judgment of the serjeant who commands; and not all the guns of a battery together.

e) Some non-commissioned officers should be appointed to observe the graze of the shot and shells, and point it out to the non-commissioned officers who command the guns.

f) Grape only should be used at night, and care taken that the gun is laid nearly horizontal.

g) Never fire without being tolerably certain of the effect of the shot. Be very sparing of your ammunition, it generally fails at the time it is most wanted.

§. 187. *Of particular Circumstances in Action.*

1) *With respect to the enemy's fire.*

a) If the batteries opposed to you are strong, you

should not open your whole fire at once, but mask part of your guns, until one party or other begins the attack; at least, they should form in rear of the troops, unlimber, and then commence firing when an opening presents itself.

b) If the shot from the enemy's battery, opposed to you, pass over and strike the cavalry, or any column, &c. that may be in the rear, do not fire again until you have changed your position.

2) *With respect to the situation of the Batteries.*

a) The batteries, placed on the flanks of the infantry, direct their fire against the enemy's cavalry; if this is defeated, the battle is in general gained, as in 1759, near Minden.

b) The centre batteries all direct their fire upon the same point, if they are not attacked. This oblique fire however is only advantageous on plains, where the shot will ricochet in front of the enemy, and occasion fear and confusion. A single break in the line often decides a battle, if there is cavalry at hand to take advantage of it; but without great rapidity, no advantage will be obtained.

If the infantry is to attack, the batteries will concentrate their fire upon the point of attack.

c) It is often advantageous to change the situation of a battery: to establish it, in the first instance, on some spot where it is not intended to remain. Every thing must be prepared for this purpose; the whole of the guns must act for some minutes, and a few must remain afterwards on the spot. By this means, the enemy may be overpowered in some part before he can receive support, and he is also misled with respect to your intentions; it is however admissible only with

6-pounders, and at the beginning of an action, when every preparation has been made for it.

3) *Advancing.*

a) If the line moves forward, the horses will be brought in front of the guns, according to §. 179 and 181; and they will advance some distance before the troops; halt and fire until the line comes up; they will continue to approach the enemy in this manner, until within 400 paces distance, and then halt.

b) If opposed by the enemy's artillery, the guns should be formed in 2 divisions, one of which should fire, while the other advances.

4) *Respecting the direction of the fire of artillery.*

a) If immediately in front of the enemy, within the distance of 500 paces, the fire of the artillery should be directed against his troops, who will probably be broken by a few rounds of grape.

b) If it is propable that the affair will soon be decided, when either party attacks, the artillery should fire at the troops. This was done by our artillery at Minden, Crefeld, and several other places; the enemy's troops abandoned their guns, which were obliged to follow, or they were lost.

c) When the enemy's guns are covered by an epaulement, &c. the artillery should direct their fire at the troops, and endeavour to break their ranks.

d) In two cases only should the artillery fire at the enemy's artillery. When they are cannonading any particular spot, pass or defile, by which our troops must advance to attack; or when the two parties remain stationary and cannonade each other, and the principal attack is expected to take place at some other point more to the right or left; it is then necessary to prevent the troops from suffering too much by the fire of the enemy's artillery.

§. 188. *Of Retreating.*

a) The waggons are first sent to the rear, and the guns then retire successively.

b) The guns, which have retired first, take up a position in the rear, and cover the retreat of the others.

c) In retreating across a plain, the guns must be formed in 2 divisions, one fires while the other retreats.

The guns retreat with drag ropes. See §. 181.

d) If compelled to abandon the guns, at least endeavour to carry off the tools and ammunition. If it is possible to spike the guns and destroy the rammer, the enemy will be prevented using them immediately.

§. 189. *Of passing a River or Defile in presence of an Enemy.*

The artillery, which is to cover the passage of the river, fires at the enemy's artillery, this only will hold him in check.

b) Part of the artillery, which opposes the passage of a river, fires at the enemy's artillery, but the principal part direct their fire against the troops, wherever they endeavour to penetrate.

PART THE THIRD.

ON ENTRENCHING. ATTACK AND DEFENCE OF FIELD-WORKS.

SECTION I.

Of the Profile and Construction of Field-Works.

CHAPTER I.

Of the Breastwork.

DESCRIPTION.

In Pl. VI. Fig. 1. *h b k n* represent a breastwork, and *d f t h* a banquette, on which the soldiers stand to fire over the breastwork. The earth for the breastwork is taken from *n o r q* by which a ditch is formed in front.

§. 190. *Height of the Breastwork.*

a) *On level Ground,* the height of the breastwork should, if possible, be 7 feet, but never less than 6 1/2. Therefore, *b a,* in Fig. 1. should be 6 1/2, or, if the time will allow, 7 feet.

b) *If there are any Hills near the work,* that is, within 800 paces, the breastwork should on that side be raised, so that the hill cannot be seen from the foot of the banquette; and it will be better if another foot can be added to this height. Therefore to determine the height of the breastwork in this case, a stake, 6 feet long, should be driven into the ground, at the foot of the banquette, *c g*, Fig. 2. that is 8 feet within *b*, the spot where the height of the breastwork is to be fixed; you then look over the stake *c g*, towards the hill, while another is fixed at *b*, the top of which you bring into the line *c* A; this second stake *a b*, will then be the height of the breastwork.

c) *On the Brow of a Hill, f c*, Fig. 3. Pl. VI. 4 1/4 feet, or less, are sufficient for the height of this breastwork. To determine the height in this instance, a person must be placed at *f*, where the greatest height is to be given; then take 8 feet further back to *k*, where a stake, *k l*, 6 feet high, is to be fixed; over which you look, and cause a second stake, *h f*, to be raised, so that you can just see, over the top of it, the highest spots in the surrounding ground, *c d i*; the height of the breastwork, *f h*, will then be determined by this second stake. If *f h*, does not exceed 4 1/4 feet, a banquette will be unnecessary.

If a breastwork is thrown up only for the purpose of covering heavy artillery, which is intended to command certain distant points, and to retire as soon as the enemy has passed them, it will not be requisite to form a ditch, the breastwork alone being sufficient for this purpose; and the earth may then be taken from within the breastwork. Thus, if the height fixed for the breastwork is 4 feet, it is not necessary that the earth thrown up should be of that height; for you are perfectly covered, if the height, *h f*, Fig. 3. of the earth thrown up, added to the depth of the excavation in the rear, (as shewn by the dotted line *e m*,) from whence the earth is taken, be together equal to 4 feet.

The height necessary for the breastwork, Fig. 4. Pl. VI. having been determined at 5 feet, and the depth of the excavation in the rear being 2 feet, the height of the earth thrown up, *b c*, should be 3 feet: part of the earth taken from the ditch may be thrown into the hollow *d*, in front.

d) *In Fig. 5. Pl. VI. a Breastwork is constructed with a River in its Front.* A ditch is here unnecessary, and the earth for the breastwork is taken from the excavation, *a b c d* behind, which being 2 feet deep, the height of the breastwork, *e f*, need not be more than 4 1/4 feet, as the men will be secure from the enemy's fire, by standing in the excavation, *a b c d*.

§. 191. *Thickness and Slope of the Breastwork.*

a) *The Thickness. b k*, Fig. 1. Pl. VI. *at the top*, should, even when the work is to be abandoned upon a serious attack, be at least 8 feet; and, when the work is to be maintained, and may consequently be exposed to a cannonade from heavy artillery, it should, when the soil is strong, be 11, and in light soil, 13 feet*.

Breastworks thrown up merely for form, or as a defence against cavalry, may be only 4 feet thick at top; but works erected at a short distance from fortresses, as a cover against heavy artillery, should have a thickness at top of 16 feet, in strong soil, and of 18, when the soil is light.

* The firing butt, used in 1792, for practising with 12-pounders, at 900 paces distance; was 9 feet thick at the top, and no shot penetrated more than 6 1/2 feet. The soil was light and sandy.

Breastworks of wood do not require a thickness of more than 4 feet to resist 12-pounders; and against musquetry, 6 inches, if deal, and 5 inches, if oak, will be sufficient. Wood, however, is attended with considerable disadvantage, when exposed to a cannonade, as the men are liable to suffer from the splinters. Masonry, 2 feet thick, will resist 12-pounders for a short time; when 4 feet thick, it affords a safe cover.

b) *The Interior Slope, b t,* Fig. 1. Pl. VI. should be as steep as possible, that the men may stand close to the breastwork, and fire over it easily. Thus its base, *t s,* or *h a,* will be about one-sixth of its height, *a b.* This slope should be always lined with sods or fascines.

c) *The Base, n l, of the Exterior Slope, k n,* when revetted with sods or fascines, should, in strong soil, be 1/4 and in light sandy soil, 1/2 of the height, *l k.* If sods or fascines are not used, and the earth is only rammed down, the base should, in strong soil be 1/2, and in light, 1/3 of the height *l k.*

d) *The Superior Slope,* (that is the summit or crest) *b k,* of the breastwork, is directed upon the outward edge of the ditch, when the work is not more than 6 1/2 feet high; but when the height is greater, it is directed about 1/2 a foot above the edge of the ditch, as shewn by *b k v.* Thus when the breastwork is upon level ground, the exterior side will be about 2 feet lower than the interior. On hills the difference is greater, and the superior slope of the breastwork often loses itself in that of the hill, see Figs. 3 and 4. In this case, the slope should be such, that not only the edge of the ditch, near *d,* Fig. 4. but also the side of the hill, to a certain extent at least, may be seen and cannonaded.

CHAPTER II.

Banquette.

§. 192.

In Fig. 1. Pl. VI. *f t*, represent a banquette, on which the infantry stand to fire over the breastwork. Its breadth *f t*, should be 5 feet, if the troops are 2 deep, but if they are in single rank, 3 feet will be sufficient. The height of the breastwork *t b*, above the banquette should be 4 1/4 Hanoverian feet, (or 4 1/2 English.) If the breastwork is 6 1/2 Hanoverian feet, the height of the banquette *t h*, or *f i*, will consequently be 2 1/4 feet A slope *d f*, should be given, when the banquette is above 1 1/2 foot high; the base *d i*, of this slope is twice its height, that the men may ascend it easily. But when *f i*, is not above 1 1/2 foot, the base *d i*, if revetted with sods, may be equal to half the height *f i*; and if not revetted, to the whole height. If there are a sufficient number of fascines, particularly when the work is enclosed, and space is of importance, no slope should be given to the banquette, and an ascent formed, where the height is above 1 1/2 foot with the fascines; this ascent is 1 foot broad, and the fascines serve as steps. Fig. 16. Pl. VI.

CHAPTER III.

Ditch.

§. 193.

a) *On level Ground, the Ditch should be at least sufficiently large* to furnish the earth requisite for the formation of the breastwork.

One of the following profiles may be adopted, as the time will allow.

1st. *In light Soil.*

1st Profile.	Breadth at top . .	10 feet.
	Breadth at bottom	2
	Depth . . . . from	7 to 9
2d Profile.	Breadth at top . .	12 feet.
	Breadth at bottom .	2
	Depth . . . . from	10 to 12
3d Profile.	Breadth at top . .	15 feet.
	Breadth at bottom .	2
	Depth . . . . from	12 to 15

2d. *In strong Soil.*

1st Profile.	Breadth at top . .	10 feet.
	Breadth at bottom.	2
	Depth . . . . from	9 to 10, according to the nature of the soil.
2d Profile.	Breadth at top . .	10 feet.
	Breadth at bottom .	2
	Depth . . . . from	12 to 15
3d Profile.	Breadth at top . .	15 feet.
	Breadth at bottom .	2
	Depth . . . from	15 to 18.

If the ditch were to be made broader, and not so deep, it would not prove so great an impediment to the enemy, as if it has the depth above mentioned. It is the depth and not the breadth that will check the enemy's attack.

If palisades are used, the bottom of the ditch must be 2 feet broad, otherwise it may be dug deeper and brought to a point.

b) The tracing, *m n*, Fig. 1. Pl. VI. of the slope of the ditch, should, when the soil is strong, be about one third, and in light soil, two thirds of the depth, *m o*.

If there is more earth than is sufficient for the breastwork, the superfluity may be thrown outwards, and spread out, as in Figs. 1 and 6, Pl. VI.

A ditch, which is several feet broad at bottom, may be considerably deepened by excavating it, as shewn by the dotted lines, *x*, Fig. 1 and 6.

CHAPTER IV.

§ 194 *Barbettes and Embrasures.*

a) *Barbette.* When the guns are not to act against particular points; when the enemy can attack in any direction, or when the work is situated on higher ground than any which he can possibly occupy, and that consequently the greater part of his shot will pass over it; in these cases the guns should be fired en barbette, that is over the breastwork: for this purpose, a mount of earth is raised behind the breastwork; the plan of which is represented by *a b f k*, Fig. 7. Pl. VI. and the profile by *a e f b*, Fig. 6.

Height. As gnns are in general about 3 feet high, the height of the mount of earth should be 3 feet, when that of the breastwork is 6 feet; and when the latter is 7 feet high, the mount should be raised tc 4 feet.

Of ascertaining the height more particularly when there are Platforms. If greater exactness is required, a height of 3 1/4 feet may be given to the breastwork above the mound of earth, for our 3-pounders 3 1/4 feet; for our 6-pounders and 5 1/2 inch howitzers 3 1/2; and 3 3/4 feet for our 12-pounders.

Height when there are no Platforms. In this case, the wheels will sink a little into the earth, and therefore the breastwork should not be more than 3 feet higher than the mount.

The Depth *a b*, Figs. 6 and 7, of the mound of earth on which the gun is placed, should be 14 feet for a 3-pounder, 15 for a 6-pounder, and 16 for a 12-pounder; but with this depth it will be necessary, that a fascine should be fastened with pickets a foot or two behind the trail, to prevent the gun running off the mound when it recoils.

The Breadth *b k* is, in general, not more than 12 feet; but it must be encreased to 16 feet, when it is required that the guns should traverse. If several guns are to be placed upon the same mound, 16 feet in breadth should be allowed for each; and in any case of necessity, never less than 12 feet.

The Rampe *d e*. The length of the ramp is twice

the height of the mound (*a e*, or *b f*, Fig. 6.) and the breadth, *e g*, Fig. 7, is 8 feet.

b) *Embrasure.* If the gun is to be directed against a certain point; or, if it is probable that artillery will be brought to act against you, from which you might suffer when firing en barbette; an embrasure should be formed in the breastwork, which must be lined on the interior side with sods, fascines or gabions; and the mound of earth is then unnecessary. Although the men are better covered, yet they cannot fire so quick through an embrasure, as the gun must be run up again into the embrasure after each shot, in consequence of the recoil. In Pl. VI. Fig. 8. the profile of an embrasure is represented by *g e h f*, and the plan by *a b c d*, in the breastwork *g i k h*.

The interior width of the Embrasure, a b should be 1 1/2 foot for 3, 6 and 12-pounders, and 2 feet for 24-pounders; and the exterior width *d c*, should be 7 feet for 3, 6 and 12-pounders, if the fire is to be directed against one point only, for instance, a bridge; but, if there are several objects at which the guns must be pointed, the exterior width must be encreased accordingly.

The Sole e i, or interior height of the bottom of the embrasure from the ground, should in general be 3 feet 3 inches: it is, however, better to give it 3 1/4 feet for 3-pounders; 3 1/2 for 6-pounders, and 5 1/2 inch howitzers, and 3 3/4 for 12-pounders; but the guns must then be placed on platforms, that the wheels may not sink into the ground.

The bottom of the Embrasure e f runs parallel with the summit of the breastwork *g h*.

The embrasures are also sometimes made from 3 to 4 feet wide within, and 10 feet without, to avoid the necessity of running the muzzle of the gun into the embrasure every time, when it is required to fire quick: in this case the men are covered from the shots in an oblique direction, but the guns cannot be fired obliquely.

CHAPTER V.

Construction of the Breastwork and Ditch.

§. 195. *Fascines for the Revetement of the interior Slope.*

Fascines should be bound at intervals of a foot, they should be from 9 to 10 inches thick *, and from 6 to 18 feet long. One man may carry two fascines of 6 feet in length; and 2 men may together carry 2 of 10 feet in length; when they are 18 feet long, 2 men will be required to carry each fascine. They are prepared upon a fascine bench, Fig. 15. Pl. VI.

The Fascine Bench consists of several crosses, which are formed by 2 stakes of 9 feet long, driven into the earth so as to cross each other at 2 feet above the ground, where they are bound together. These crosses are fixed at 2 feet asunder; they are placed in a line and form so many forks, upon which the brushwood is laid.

Pickets are driven into the ground at each end of the bench, about half a foot from the extreme crosses, to

* When a foot thick, they require a great deal of brushwood, and are heavy and difficult to handle.

regulate the length of the fascine. These pickets must not reach quite so high as the fascines, that the projecting part of the brushwood may not be bent by them.

The crosses for a good fascine bench must be formed of straight stakes 2 inches thick; they must be in an exact line, and of an equal height. In constructing the fascine bench, a mason's level or a measuring rod should be made use of, which being laid occasionally on the crosses, any irregularity may be perceived and remedied.

Distribution of the Workmen. Four men are allowed to every fascine bench; 1 of whom writhes the twigs, and the other 3 place the brushwood upon the bench and bind the fascine; which is performed in the following manner: 2 men draw the brushwood close together by means of a choke, which is a rope or chain with a stick at each end, Fig. 15. A; the third man then binds it firmly together with a twig. Fascines may also be made without using the choke, by merely binding the brushwood as tightly as possible; this is generally the case in the field, where expedition is required. An instrument should be prepared, Fig. 15, B. with three pieces of wood, forming three sides of a square, each 10 inches, for the purpose of finding if the fascines are of the proper thickness, after 2 or 3 twigs have been bound: if the fascine fits into the instrument, it has the proper thickness. If you are not provided with this instrument, a rope of 30 inches in length may be passed round the fascine, which will prove whether it has the proper thickness; if the rope hangs loose, more brushwood must be added, or the fascine must not be drawn quite so tight: however, it is better that it should be bound as closely as possible, as the brushwood is apt to loosen when it becomes dry.

When the fascines have been bound, the brushwood, that projects over the pickets, at the ends of the bench, should be sawed off.

If more than 4 men are placed at a bench, they will be in each other's way. When lighter and more pliable fascines are required, 2 men should lop the small branches off the brushwood, and 2 or 4 more should be employed in cutting and carrying it.

Tools and Workmen for each Fascine Bench. To every fascine bench either from 4 to 6, or from 8 to 10 workmen are allowed: the tools are, 1 hatchet, 1 fascine knife, a choke, a saw, a gauge and a measuring rod, in the first instance; but in the second, 2 more hatchets must be added; and in the third instance, 4 or 6 hatchets besides those above mentioned.

Fascines of a superior kind. When these are required, it is necessary 1) That the brushwood should be free from leaves, and that the strongest should be placed in the middle. 2) That in binding them, all the knots should be on the same side, that they may be laid towards the earth; and also that the ends of the twigs should be fastened into the nearest bands. 3) That after it is bound, the fascine should be cleared with the fascine knife; and 4) That the ends are bent inwards and finished neatly.

Good fascines should be equally thick, round and straight, and show no ragged branches.

Fascines used for lining embrasures, should be made of strong brushwood; if they are thin and leafy, they are soon destroyed by the gunpowder.

Number of Fascines that may be made at each Bench in a Day. In general, 200, and with very good workmen, from 3 to 400 feet of fascines may be completed at each bench in a day. *

Pickets. A picket 1 1/2 inch thick, must be allowed for every 4 feet of fascine. These pickets are driven through the fascines, when they are used for the revetement of slopes. One half of the pickets should be 6 feet, and the other half from 2 to 4 feet in length: in addition to this, an understake 6 feet long will be required to from 6 to 10 feet of fascine.

The Number of Fascines required, may be found by multiplying the number necessary for the height of the breastwork, above the banquette, by its length in feet. Thus, if the height of the breastwork above the banquette is 4 feet, and the length 200 feet, 5 fascines must be placed one above the other for the height; then 5 multiplied by 200 gives 1000 feet, the length of fascine required.

---

* Our recruits in 1792, were able, after having been trained only 3 days to the work, to complete 300 feet, although only 4 men were allowed to each bench.

CHAPTER VI.

Of the Time necessary for the Construction of a Fieldwork—Workmen—Entrenching Tools.

§. 196. *Workmen and Time.*

A fieldwork of a weak profile may be constructed in eight hours, if 4 workmen are allowed to every pace of the centre length of the ditch; half the number being provided with entrenching tools, so that one half will be constantly at work, and will be relieved every two hours. A fieldwork of a strong profile requires double the time for its construction.* It is, however, to be understood, that the interior slope is revetted to the height of 2 or 2 1/2 feet above the banquette with fascines ready prepared, or with inferior hurdles, the materials for which are to be brought in waggons by other people.

The revetement with sods, if merely for the interior slope and the banquette, requires half a day more, and if for the exterior slope also, nearly a whole day; particularly if the sods must be procured at a distance.

§. 197. *Entrenching Tools.*

One spade must be provided for each workman (as

* One man may excavate 200 cubic feet of earth in a day, and also throw it up 2 or 3 feet on on side; but as, in the construction of a fieldwork, the man who excavates the earth, can seldom throw it at once to the spot where it should lay, 2 men must be allowed to 200 cubic feet, or, generally, 1 man to 100 cubic feet of earth.

some will soon be lost or destroyed) 1 pickaxe must be allowed to from 4 to 8 men, for the purpose of loosening the earth when hard; 1 large mallet to every 20 men, with some hatchets, laths, nails and tracing lines, or ropes of twisted straw, some stakes and pickets, and at least 1 measuring rod. It is often difficult in war to procure entrenching tools fit for use; therefore, when a work is undertaken, particular attention must be paid to them. They are either furnished from the park of artillery, or taken from the country people, and sometimes those with the guns are made use of. It is of consequence that grindstones should be provided, for sharpening the shovels, &c. When a fieldwork is to be constructed very accurately, mason's levels with lines and plummets will be requisite.

§. 198. *Of Tracing the Breastwork.*

The breastwork is traced by means of a cord, chain, gun-match or straw band. The line *a f*, Fig. 10. Pl. VI. having been determined and marked by stakes, the cord is stretched along the line from stake to stake, and the ground notched on both sides with a spade; perpendiculars *a d*, *e h*, &c, are then raised, by the eye, upon this line, on which the lower breadth or base *a c* of the breastwork, and the upper breadth *c d*, of the ditch are marked. This should be repeated upon all the perpendiculars *e f*, &c; after which, the cord is stretched between *c g*, and *d h*, &c. and the ground being notched again as before, the principal lines will be determined.

§. 199. *Division, Placing, and Instruction of the Workmen.*

The ditch should be divided into parts of 6 or 12 paces each, and the workmen formed in as many squads as there are of these divisions: each squad will be under the command of a non-commissioned officer, and will be stationed to one of the divisions of the ditch: without this arrangement, confusion would soon take place.

If there are more than 2 workmen to each pace, the squad should be subdivided, and the men work alternately. Before the work is commenced, the non-commissioned officers should be assembled and instructed in the manner in which it is to be performed; the tracing of the upper breadth of the ditch should be pointed out to them, and they should, likewise, be made acquainted with the breadth at bottom, and the depth; being acquainted with this, they will be able to proceed by themselves; and if the breadth of the bottom is stated to them as less than is really intended, or, if the excavation is begun about a foot from the interior line of the ditch, any errors can be easily corrected afterwards, by removing the superfluous earth. In order to guide the workmen; profiles of laths should be formed at several points on the line of the work. See Fig. 17, Pl. VI. *

---

* A plummet and measuring rod will be necessary to erect this profile accurately. For instance, if the base *b c*, of the slope is to be 2 feet, the height being 4 feet; the lath *a b*, should be first fixed perpendicularly in the ground; and to ascertain that it is perfectly perpendicular, a plummet should be applied to it, or a stone should be dropped down along it; and when it is found to be true, the lath *a c*, should be fixed so that *a b*, will be 4 feet, and *b c*, 2 feet.

§. 200. *Actual Construction of the Breastwork.*

When the breastwork is the height of the banquette, it will be necessary to commence the interior revetement, either with sods or fascines. The sods should be cut about [illegible] inches thick, 1 1/2 foot long, and 1 foot broad; they should be laid in the same manner as bricks in a wall (so that the parts where the sods join, are not immediately over each other) and each sod should be fastened with a picket 1 1/4 foot long.*

If there is not time sufficient to allow of the slope being revetted with sods, and that fascines are to be used, you begin by fastening the first fascine to the ground, in a small trench which is dug to receive it, with pickets 2 1/2 feet long, at intervals of 4 feet: the second is then laid upon it, in such manner that the bands are placed exactly over those of the lower one, and

---

Attention should be paid that the profiles are in a direct line.

In order that the summit of the breastwork may be level; the profiles should first be erected at the angles, and the intermediate ones regulated by them. When more permanent works are erected before a fortress, the line should be levelled in order that the summit of the work may be perfectly horizontal.

* They should be placed so that by cutting a small part away on the interior side, the exact slope may be obtained; this is best done by stretching a line from one profile to the other, and laying the sods along it. If it is meant that they should lay very close, the ends where they join should be cut even, and they should be pressed together. As the revetement proceeds, care must be taken to throw up the earth behind them.

When the sods are dry, they should be moistened, and thus should always be trod or beat well together. After each row is finished, the uneven earth should be removed, that the next may be laid with exactness. When the soil is sandy, thin twigs should be laid between the sods, and should extend some way into the breastwork, to enable the sods to resist the pressure of the sand.

the knots always turned within, towards the earth: this should also be picketted at intervals of 4 feet. The same method is to be pursued with the remainder, taking care however, that the ends do not come exactly over each other, but that every fascine rests upon 2 in the row beneath. When the revetement is thus carried to the height of 3 feet, the fascines (unless the soil is very stiff) uld be anchored, that is, each fascine is fastened with twigs to stakes driven into the breastwork at 6 feet from the fascines, and at intervals of 6 feet from each other. The earth forming the hreastwork, should, from time to time, be trodden or beat together. *

The breastwork (Pl. VI. Fig. 16.) is revetted within with fascines, and without with sods. If the time is short, and brushwood not to be got, the interior slope may be lined with hurdles, which should also be anchored. ** But then the banquette only should be completely revetted, and the breastwork not higher than from 3 to 3 1/2 feet above the banquette.

---

* To preserve the slope exact, a cord should be stretched from one profile to another. It is very essential that the fascines should be laid accurately before the stakes are driven in, as you are very apt, without great care, to drive them in perpendicularly.

Lastly, strong stakes should be driven in from top to bottom, and anchored strongly; the method of securing the *anchor-stakes* is, to place them horizontally, and drive in two stakes to the right and left within them, (Fig. 16. Pl. VI.) this, however, is only necessary when the soil is very light, or particularly when it is sandy, which will always force the fascines out a little, so that the slope will not be so great as it was at first traced.

** It is a false idea, that fascines are not proper for the exterior slope; the firing butt which we erected for practice, was revetted with fascines, and the earth did not fall so much as in those where sods were used. All the stakes will not be shot away, and those that remain, will be sufficient to support the fascines and the earth.

§. 201. *Method of tracing Fieldworks.*

A Redoubt. *a b e f* (Pl. VI. Fig. 11.) is traced in the following manner: the side *a b*, is first laid down as pointed out in §. 198. upon which the 2 perpendiculars, *a e* and *b f*, are then raised by the eye, and made equal to *a b*, by which the side *e f*, is also determined; if this side, however, proves not to be so long as *a b*, it should be made so, and the direction of the lines *a e*, and *b f*, altered so as to meet the ends of *e f*.*

A Flèche. *a b e*, (Fig. 12. Pl. VI.) is traced in the following manner: a cord is divided into three equal parts, and each part being stretched out, a triangle *d b e*, will be formed; each angle of which will contain 60 degrees: the sides *b d*, and *b e*, should then be prolonged to the length intended to be given to the faces of the flèche.**

A Hexagonal Star Fort is traced as follows: a circle should be described on the ground from the point *c* as a centre (Fig. 13. Pl. VI.) and then divided into 6 parts, by carrying the radius six times round the circumference. Having thus obtained the points *a b g f e d*, the line *a c* should be prolonged to the length intended to be given to the face of the redoubt; if this length is to be 40 paces, *h c* should be made equal to 40 paces; and the same method is to be pursued with respect to *c b*, &c.

---

* If greater accuracy is required, the Perpendiculars upon *a b* may be raised as follows: take 3 paces from *b* to *c*, and then determine a point *d*, at 4 paces from *b*, and 5 from *c*----*b d* will then be perpendicular to *a b*.

** If the saliant angle of the flèche is to be cut off, that a gun may be placed there, 10 feet should be taken on each face from *b* towards *a* and *c*, the points *d* and *e* being thus determined, will give the flèche *a d e c* as required.

By this means, the points *h i m*, &c. are gained; then upon *h i* let the equilateral triangle *h k i*, be constructed: this may be done by makiug first a small equilateral triangle, *h o p*, with a cord divided into three equal parts: the side *h o* is then to be prolonged to *k*, and *h k* made equal to *h i*; by which the line *k i* will also be determined. The lines *i l m*, &c. are obtained in the same manner, and thus the tracing of the redoubt *h k i l m q n*, &c. is completed. Here the salient angles at *k*, *l*, &c. each contain 60 degrees, and the re-entering angles at *h*, *i*, *m*, &c. each 120 degrees.

A star fort of 5 or 7 *points* may be traced in the same manner, observing, however, that the circle *a b g*, &c. should be divided into 5 or 7 equal parts by trials.

SECTION II.

Of the Impediments, which should be presented to the Enemy's Attack of a Redoubt or Entrenchment.

§. 202.

When there is nothing but a common ditch in front, and no other obstacle exists against an attack, the bravest troops will be lost in a fieldwork;* therefore impediments should be prepared, within reach of the fire, and if possible, near the work. This may be done :

a) By *Pallisades*, or strong stakes of wood, having one end pointed, and fixed in the ground, so as to prevent the enemy passing the ditch without leaping over them. They answer this purpose, when placed obliquely and immediately in front of the ditch, the points being turned towards the enemy, and elevated about 3 1/2 feet above the ground, as in Pl. VI. Fig. 9, *h*; the troops can then fire over them, though they still cover the enemy in some measure, and he may destroy them with his guns. When the pallisades are placed in the ditch,

* For although they are covered, the enemy has in turn other advantages. He previously cannonades the work with his guns or howitzers; he places his confidence in the assault, and nothing will prevent its succeeding: the troops in the work rely entirely on their fire, which, however, when it is scattered as in this case (the space covered by the enemy being greater than the redoubt) seldom checks his advance.

as in Fig. 6, it is impossible for the enemy to leap over them, they afford him no cover against the fire from the work, and he cannot destroy them with his cannon.*

b) *Fraises.* These serve to prevent the enemy from climbing up the breastwork, they are represented by *m n*, Fig. 6.** they are not so much to be depended

---

* When placed in this position, the points should be of equal height with the crest of the small glacis *l o m*, which is from 1 to 1 1/2 feet high; from the point *h*, 3 feet below, the exterior slope of the ditch runs to *l*, so that *l k* is about 2 1/2 feet: beneath this point, from *h* to *p*, (between 5 and 6 feet) the ditch has no slope, and the earth is supported by the pallisades. The greater the thickness of the pallisades, the less danger there is of their being cut down. It is found very difficult to cut down stakes 3 inches thick and 4 inches broad, when fixed in this position. They should not be more than 2 or 3 inches asunder, so that a person may not be able to step upon the rail, which connects them together. When in this position, they need not be sunk above 1 1/2 foot below *p*; but then they should be nailed to a rail, 4 or 5 inches thick, placed on a level with the bottom of the ditch; above which, (at 2 1/2 feet below the top) they are connected by a ground-sill rail (at *q*) 8 inches thick, to which they are fastened with strong nails. Supports are fixed behind the rail, as shewn at *q*, to prevent the enemy from forcing the pallisades backward. The pallisades should be well pointed at the tops; their length depends on the depth of the ditch; if *l p* is 8 feet, they should be at least 9 feet long.

By this arrangement, the depth of the ditch and the pallisades unite to obstruct the enemy's advance, when under the heaviest fire from the work: if the ditch was shallow, and the pallisades placed in the middle of it, both together would not impede the enemy's advance when exposed to our fire, and he would be able to cut down the pallisades under cover. When pallisades, are sunk obliquely in the ditch, in the direction of the line *p y* Fig. 1. the enemy can slide down the slope of the ditch, or leap over them.

One man may make about 30 pallisades in a day. In sinking them into the ground, a trench should first be dug, and 2 pallisades placed accurately, at 20 or 30 paces from each other; between these a cord is stretched, by which the others are fixed. About 30 pallisades may be fixed by 1 man in a day.

** Fraises are similar to pallisades in regard to their size and distance

on as pallisades, as they may be pulled down, under cover from our fire.

c) *Trous-de-loup,* are placed in front of the ditch, and present a formidable obstacle to the enemy, particularly if a small stake, pointed at the top, is driven in the middle of each; but if it is sandy ground they are easily passed. They are in 2 or 3 rows, near each other, in front of the ditch; the second row is disposed in such a manner that each trous-de-loup covers the space between 2 in the first row; and the same disposition is observed with the third.

Three rows of circular trous-de-loup are represented to the right in Fig. 18. Pl. VI. and to the left 3 rows of square ones. They are shewn in profile in Fig. 19.*

d) *Abbatis.* When a fieldwork is situated in, or immediately in front of a wood, all the trees round it should be cut down, by which a natural abbatis is formed. If a work is placed near a wood, trees may be drawn by horses to the edge of the ditch**.

---

from each other; they are secured by 2 beams at *n* and *m*, and are laid with their points inclining downwards, that they may not be so much exposed to the enemy's guns.

* The ditch is represented at A. *f e* = 6 feet, *e d* = 5 feet, *e c* = 3 feet, *a c* = 8 feet. In fig. 18, *p r* = 12 feet, *k b* = 6 feet, A *l* = 6 feet, A *d* = *d q* = *l n* = 6 feet.

Trous-de-loup are particularly made use of in front of works, which are defended by others, as pallisades cannot there be employed with so much advantage, as the case shot, fired from the other works at the enemy when on the edge of the ditch may injure the garrison; in this case therefore, the trous-de-loup are 40 or 50 paces from the ditch, otherwise they are immediately in front of it. Trous-de-loup are also generally placed between two fieldworks, to prevent the enemy from passing. One man may complete 3 trous-de-loup in a day.

** They are laid with the trunks towards the works, and the branches to-

e) *Thorn-bushes and Harrows.* When there is not sufficient time to construct trous-de-loup or abbatis, thorn-bushes placed on the edge of the ditch are very useful, particularly if iron harrows or planks, with large iron nails driven into them, are placed underneath; indeed, when there are no pallisades, these are almost indispensable.

§. 203. *Continuation.*

f) *Fougasses and Shells.* When a box *a*, Fig. 9. Pl. VI. is buried in the ground at a certain distance from a fieldwork, and a small channel or trough, *ff*, provided with a saucisson, is carried from it under ground into the work, and the earth afterwards levelled over it; the enemy, when he arrives at the spot *d*, may be blown up, by setting fire to the saucisson, and thus inflaming the powder in the box, that is, by springing the mine.

The space, in which the box *a*, containing the powder, is lodged, is called the *chamber*; the hole *d e*, which is dug for the purpose of placing the box under the earth, is called the *well*; the fire is communicated by the saucisson through *e c f f*, the cavity *f a g*, formed by springing the mine, is called the *funnel*, and the depth *d a*, the *line of least resistance*.

Shells are sometimes used instead of mines, and when 2 or 3, filled with powder, are buried near together, at the depth of 5 or 6 feet, they will produce the

---

wards the enemy, and so close together, as to render it almost impossible to pass between them. Sometimes the trunks of two trees are placed across each other; but the principal object is, that they should be picketted down strongly, so as not to be easily removed.

same effect. They are of great advantage in the ditch, when placed singly at the distance of 10 feet from each other, particularly at the angles. They must also have their troughs and saucissons.

The well should be about 4 feet square.

The hole, through which the saucisson passes into the chamber, should be 1/7 of the depth of the chamber.

The following charges are to be made use of: When the line of least resistance is

Feet.		lb.	oz. of powder.
3	- - -	2	8
4	- - -	6	0
5	- - -	11	11
6	- - -	20	3
7	- - -	32	2
8	- - -	48	0
9	- - -	68	5
10	- - -	93	12
11	- - -	124	0
12	- - -	162	0*

* With this charge, the diameter *g f* of the funnel will be double the line of least resistance; with a greater quantity of powder, it will be larger.

The inside of the box, containing the powder, is nearly 1/8 of the line of least resistance: the form of the box is cubic, and it is pitched over within. The saucisson is from 1/3 to 1/2 an inch thick, and is made of fustian or linen. The trough is formed of boards, and should be just large enough to receive the saucisson, which is nailed to it. When the soil is strong, the trough should be placed 1 foot, otherwise 1 1/2 foot under ground. When there are several troughs in the same ditch there should be at least 1 foot of earth between them. The well must be completely filled with earth, well rammed together, etc. When several mines are formed near together, the distance between the chambers should be double the length of the line of least resistance.

Fougasses are formed at from 10 to 13 paces from the ditch, at the point where the enemy is most likely to advance, and in front of the weakest

g) *Inundations.* When there is a brook or river, which is fordable, in front or on the flank of a field-work, the water should be stopped below, by shutting the floodgates at a mill, until it has risen to such a height as to become impassable. Or if there is no mill near, it may be effected by means of timber, thrown into the river behind a bridge, or fastened to trees; then a quantity of more timber, straw and brushwood, being immediately heaped up behind it, a dam will be formed.

If there is sufficient time, a dam may also be formed across a brook or river, and a sluice constructed in it, by which means the depth of water may be increased at pleasure*.

---

parts of the work; at the angles, (as shown in Fig. 20. Pl. VI. where the chambers are at *a*, *b* and *c*, and the saucissons at *d* and *e*) but particularly in defiles; also at spots, near steep hills, which cannot be seen from the top; and they are used to the greatest advantage with works, that it is necessary to abandon on a serious attack. The worst circumstance attending fougasses is, that they seldom spring at the moment the enemy is immediately upon them; unless therefore he is obstructed by thorn bushes pallisades, etc., it is only accidentally that he is destroyed by this means.

* In Fig. 14. Pl. VI. a dam with a small sluice is represented, to the left of which at A is the same in profile. The pressure of the water is against the side A g, and A *e* is the upper breadth. Here the height of the dam, to which the base of the slope is always equal, is taken at 9 feet. The breadth of the dam at top should never be less than 9 feet, whatever the height may be, in order to allow the waggons, necessary in the construction, to move upon it.

The channel of the sluice *a a* Fig. 14. is wider at the ends than in the centre, and on its sides stakes are driven in at *b e b* and *b e b*: there is likewise a row of stakes in the middle of the dam *e e*. In the revetement *b e b* and *b e b*, two stakes are left out at *e* and *e*, and into these intervals the floodgate, marked *c* in Fig. B, is introduced, and is moved by means of a lever, as there represented. The gate of the sluice is first shut down, and may then be raised according as it is thought necessary to encrease or diminish the quantity of the water. It will be difficult to accomplish this work without the assistance of a miller and a carpenter. The dam may also be carried farther to the right and left, in order to prevent the water running off so soon, and to extend the in-

§. 204. *Choice of Impediments.*

With single works, (particularly if they have no flank defence) pallisades should first be formed, and then trous-de-loup and abbatis; but when a work is defended by others, trous-de-loup should be made first, and the pallisades afterwards, for the reasons alledged in §. 202.

If a fieldwork cannot be completed in sufficient time, you may, at the same time you commence its construction, cause a barricade, consisting of harrows, thorn-bushes, waggons, wheels, trees, &c. to be formed round the work; ditches may likewise be dug, and stakes driven into the ground obliquely, with the points inclined towards the enemy.

Fougasses and inundations possess the advantage of keeping the enemy at a distance, independent of the fire from the work. Fougasses besides create a general panic; they are however only useful at places where the enemy is checked by pallisades, thorn-bushes, &c. Fougasses and inundations should therefore be employed in every possible case: the preparations for the latter, should be begun at the same time with the work; but the former should only be constructed, when the work, together with the pallisades, trous-de-loup and abbatis, are all completed.

---

undation. Pits or ditches may be dug before the inundation takes place, at those points where it is supposed the enemy will attempt to cross, and also to render the passage more difficult, if the water is not of sufficient depth. When the sluice is only for a temporary purpose; thick boards, whose length should be equal to the breadth of the channel, may be slid into the spaces between the stakes at *e e*, in order that the channel may be closed and a small quantity of water only pass through. Great attention must be paid to the quantity of water, breadth of the channel, etc.

SECTION III.

Of single Fieldworks, Block and Guard-Houses, and Entrenchments.

§. 205. *Size.*

When a work is to be constructed for 300 men, and the garrison, when drawn up for it's defence, is, as usual, to be formed 2 deep, the circumference must be 300 feet; 2 feet being allowed for each file; the interior side of the breastwork, along which the troops stand, must therefore be 300 feet in length.

When the garrison is to dwell within a square redoubt, and is to be formed 2 deep for its defence, each side of the work must be 100 feet in length. For instance, if each side were only 75 feet; the 300 men, required to man the work, would not have sufficient room to encamp within it; still less can they, when the redoubt is smaller. If the breastwork is to be manned by a single rank, 130 men may dwell conveniently in a square redoubt, the sides of which are 60 feet in length*. Consequently there is not sufficient space for the garrison to encamp within small forts.

* Each soldier requires 31 square feet, including the space between the tents. One tent, 11 feet long, and the same breadth is allowed to 5, or at the utmost, 6 men. In a block-house, only 16 square feet are allowed to each man.

§. 206. *Form of single Fieldworks.*

a) No form is more advantageous for fieldworks, whose circumference does not exceed 500 feet, than the common square redoubt, which is represented by *a e f b*, Fig. 11, Pl. VI. The angles however of such redoubts should be cut off, as shewn here by the line *c d*; but this line, *c d*, should never exceed 8 feet in small, and 16 feet in large redoubts.

b) When the circumference exceeds 500 feet, there is no form better than what is called the cross redoubt. It consists of 4 half redoubts, Fig. 2, Pl. V. Here the angles are likewise cut off*.

§. 207. *Of Block and Guard-Houses.*

a) *A small guard*, which is posted at a distance from the quarters, &c. may be easily surprised, particularly in a wood; and besides which, it cannot remain in the open air during the winter: for these reasons, guard-houses are necessary, which may be able to resist slight attacks. The plan of a guard-house of this kind is represented in Fig. 12, and the profile in Fig. 13,

---

* If single redoubts are merely furnished with pallisades; 2 rows of trous-de-loup, or an abbatis, should be added in front of the angles: when these are placed entirely round the work, greater attention must be paid to the formation of the trous-de-loup. The same measures that are taken in front of the angles of common works, will also be necessary before the most exposed sides (for instance, at *b*), of the cross redoubts. It is to be understood, that the angles and sides should undergo some alteration, according to the nature of the ground.

The truth of the rules here laid down is demonstrated in my Manuel for Officers, chapter 6, section I. Vol. II. and approved of by all competent judges.

Pl. V. It is 15 feet square, and is sufficiently large to contain a garrison of 30 men*.

b) When a guard-house is intended to resist a powerful attack, or heavy artillery, for several hours, a building should be constructed according to the profile represented in Fig. 5, which is 20 feet square; it is calculated for 30 men, and is provided with a double flooring of timber**.

---

* It is built with posts, 20 feet high, and 9 or 10 inches thick. The earth is excavated, about half a foot in the centre, for the purpose of forming a banquette, which should be 1 1/2 foot high. A floor is formed at 8 feet above the ground, and another at 7 feet above that. The angle posts are connected at top by a plate *k*. Both floors are also supported by plates, *f* and *e*, as shewn in the profile, which are laid upon strong posts, as represented in the plan by *a b* and *c*. On each side are 4 loopholes, which are formed 6 feet above the ground, that the enemy may not fire through them; they are 8 inches wide within, and 4 without. A wedge is attached to each loophole, that it may be closed or opened, as required. A quantity of stones and a supply of water, are prepared on the upper floor; the former for the purpose of being thrown on the enemy, and the latter in case of fire.

Thirty men could not erect a fieldwork, which would hold out a moment against a night attack; it would also require great labour, and would be inconvenient for the guard.

** The earth from the space *d*, where the guard-house is to be erected, should be excavated to the depth of 4 feet, and thrown outwards; after which 2 rows of posts 1 1/2 feet thick, should be sunk close to each other on the circumference; the interior row should be 9 feet, and the exterior, 13 feet high above the ground; and at 2 feet above the ground, loopholes should be formed. Joists, *b*, 1 1/2 feet thick, should be laid upon the inner row of posts; and others of the same strength, are placed over them crossways; and above this earth is heaped to the depth of 3 feet, as shewn at *a*. A ditch 4 1/2 feet deep should be dug on the outside, and the earth thrown up 2 feet towards the field. The posts are sunk 2 or 3 feet into the ground, to obtain the necessary solidity; the inner row is fastened to the joists which rest upon them, and the outer row are also connected to the joists; this is not shewn in the plan. Thirty men may defend a guard-house of this kind against any attempt at escalade, if a sufficient supply of stones are collected at the top,

c) When a guard-house is to withstand an attack, even of heavy artillery, for 24 hours, or longer; the method of Captain Muller, of the Prussian engineers, should be adopted for the construction; long beams should be used, the ends of which should be halved and let into each other; the 4 beams then form a square, upon which another of the same construction is placed; this is repeated until the necessary height is attained, and thus a wall of beams is formed.

When the wall is 6 or 7 feet high, the top should be covered with joists laid close to each other across the building, over which a sufficient quantity of earth should be thrown, and should likewise be heaped up against the side-walls, in which loopholes are cut, and the whole building is surrounded with a ditch.

In Pl. V. Figs. 15 and 16, half of a building of the above kind is represented in profile*; but as such a

---

The lower loopholes are 1 foot wide on the interior, and 4 inches on the exterior side; they may be closed, if necessary, by wedges. This guard-house is constructed sooner than a field-work of equal size, particularly if the latter is to be provided with pallisades, and yet the lower part of the building is tolerably secure from shells or shot.

* In Fig. 14, the joists are cut lengthways, and in Fig. 15, they are cut across. In this instance, there are 5 beams, which are placed one upon another, and give a height of 7 feet. Each side of the block-house is 24 feet in length within. The loophole is cut between the 3d and 4th beam, its breadth without is 4 inches, and within 1 foot, and its horizontal length is on the outside 19, and on the inside 24 inches. One foot of the beam should be allowed to remain in the centre of the loophole, as shewn at *a*, for the purpose of supporting the upper beams. The banquette *c* is 2 feet broad, and the remainder of the earth within is excavated at *d*, to the depth of 4 feet. The side beams used for the walls, should be smoothed on 2 sides, and the joists in the roof on 3 sides. At each end and in the centre, another beam, *e*, is laid over the joists, and fastened to them with strong wooden pins. The crevices between the beams should be stopped with moss and stiff clay, which should be heaped up to the height of 1 foot in the centre, and 2 or 3 inches

building may still be taken by a powerful attack, as it is not defended by a flanking fire, and as the escalade is easy; it would perhaps be better, to carry the walls up higher by means of beams on the outside, so as to form an open space at top, from whence shells, stones, &c., may be thrown upon the enemy, when he reaches the ditch.

The blockhouse, Fig. 1, Pl. V, is constructed nearly in the above manner, and was actually erected near Schwedelsdorf, in 1778, but nevertheless it was taken.*

---

towards the sides; above the clay half a foot of dung, and over that 1 foot of earth should be placed. The thickest part of the earth (*f*) is from 8 to 12 feet. The ditch is from 12 to 14 feet broad, and 8 feet deep.

If the building is required to have more strength, 3 rows of beams may be laid close to each other.

The garrison for this blockhouse will be 20 men.

* It may be observed in addition, that the sides might be constructed with 3 walls of beams, for the blockhouse at Schwedelsdorf, which was formed with two, could not resist the Austrian artillery. The pallisades, described in section 202, would be more advantageous than those represented in Fig. 1.

The earth, at top, should be 4 feet thick, which is the greatest depth, that shells from field howitzers of the largest calibre, ever penetrate. The vertical length of the loopholes in the blockhouse at Schwedelsdorf was 1 1/2 foot, and the breadth horizontally was on the outside 4, and on the inside 10 inches.

If required to erect a blockhouse for 2 or 3000 men, the plain form of a square should still be preserved; and the roof should be supported with one pillar on each side, that is 40 feet in length, and with two when 60 feet in length. In this work there would be a smaller circumference to defend, and there would be a sufficient space at top, from whence any attempt at escalade might be repulsed by the men with their bayonets and with shells and stones. If there are three walls of beams, the two exterior of which are close together, and that the thickness is every where 3 feet; light artillery will have no effect against such a building. A supply of water in case of fire, is an object of great importance, as the building can only be destroyed by red hot shot, large shells or a great number of heavy guns.

§. 208. *Direction of the Lines and Works of extensive Entrenchments.**

a) When the lines are only to be occupied by infantry; an entrenchment composed of salient and re-entering angles, as shewn in Fig. 24, Pl. VI, is the best: the re-entering angles *a b c* and *c d e*, should not exceed 120 nor be less than 90 degrees; and the faces *a b*, *b c*, *c d*, &c., should be as nearly as possible 150 paces.

b) When the lines are to be defended by artillery, the faces *a b*, *b c*, *c d*, &c., may be 300 feet long when 3-pounders, and 500 when 12 pounders are employed. Above this distance no artillery will be able to defend a work.

c) When, as is generally the case, the works are defended by infantry and artillery together, both the above methods should be combined, as shown in Fig. 10, Pl. V. The guns in the work *a b*, defend the line *c e*, which is 500 paces long; and the fire of the infantry at *c*, *g* and *h* flanks the lines *c g*, *g h*, and *h e*. **

---

* That the following arrangements are superior to any now known, is demonstrated in the second Volume of my Manuel for officers.

** A more particular description of the above works. Suppose *a d*, Fig. 10, Pl. V. to be part of the line which is to be fortified. At *a*, let half a square redoubt be constructed, the exterior side of which should not exceed 15 paces, on the extremities of this to the right aud left perpendiculars, *c e*, 500 paces in length, should be raised, and at *g* and *h* flanks, 40 paces long, should be formed, so that *c g* may be =200, *g h* =175 and *h e* =125 paces. The side of the redoubt will be 80 paces. By this arrangement, the redoubts will be 850 paces from each other, and between them two broken lines, as shewn at *c e*.

The lines *c g*, *g h*, and *h e*, have a ditch furnished with pallisades, and there are three rows of trous-de-loup in front of the whole extent, at about 40 paces from the ditch. The guns which are placed in the redoubt defend

d) When a space of ground is to be fortified by single works; square redoubts should be disposed, at 500 paces from each other, in the manner represented in Fig. 11, Pl. V. by Nos. I, III, and IV; the angles towards the enemy should be cut off. The sides of the redoubts should be about 150 paces long, so that a batallion, formed 2 deep will be able to man one of them. Trous-de-loup should be formed at between 30 and 40 paces in front of the redoubts, as shewn in the Fig. The ditch should be pallisaded; and between each two redoubts, 2 rows of trous-de-loup should be placed more in front and extending nearly the whole distance between the redoubts.*

e) For the rest, it is to be understood, that the entrenchments, shewn by *a b c* and *d*, Fig. 11, are subject to great alterations, according to the circumstances of the ground; that the lines are sometimes more and sometimes less extensive, and that the works are placed closer or more distant as a height, &c., may render it necessary.

---

the whole line *c e*, and take the enemy in flank, and the infantry stationed at the lines *c g*, *g b* and *b e*, defend each other by a cross fire, while he is passing the trous-de-loup, and when he arrives at the ditch. If the work *a b* is closed with pallisades, formed as directed in section 218, it may be defended until the enemy, having penetrated any part of the line, shall be driven back.

* If it is intended to encrease the number of these works, a second row of redoubts may be erected a little in the rear of the intervals between the former; so, as to enable them to fire with grape from their front sides along the redoubts of the first row without hurting the men in the redoubts. The prolongation of the front sides of the redoubts in the first line, should therefore fall about 20 or 30 paces in rear of the works in the second: the second line has trous-de-loup immediately in front. In an entrenchment of this kind, infantry, as well as artillery, may make a most powerful defence; besides, each redoubt is in itself of considerable strength, and there is sufficient space between them for the troops to act. In both rows, the rear sides of the redoubts need only be closed with pallisades, the formation of which should be similar to those used for a tambour, section 218. By this much labour will be saved, but on the other hand, so good a fire will not be gained to the rear upon the enemy, when he may have penetrated through the first line.

§. 209 *Entrances, Bridges and Traverses.*

a) *Entrances.* In field-works, were guns are not used, the entrance should be 4 feet wide, and should be carried through the breast work either obliquely or in a serpentine form. When there are guns the entrance must be 8 feet wide. The entrance should be closed with chevaux-de-frise or pallisades, and about 12 feet behind it a straight breastwork, 6 feet thick at the top, should be thrown up. The entrances into entrenchments of greater extent, must be at least 9 feet wide; but if the troops are to move through it during an attack, from 20 to 40 paces will be necessary. It will be well also to dig a ditch in front of the entrance, and to lay planks or spars across, which may be thrown into the ditch in case of an attack or an alarm during the night.

b) *Traverses.* In consequence of the general use of heavy artillery, and particularly on account of the howitzer shells, traverses are necessary in every field-work of any importance. They are small mounds of earth 6 feet high, 8 feet thick at the bottom, and 6 feet at the top, and are from 10 to 12 feet in length. They should be revetted all round with hurdles, which are fastened together at the ends with strong twigs, for the purpose of supporting the earth.

In the redoubt *a b c d*, are traverses at *e*, *e*, *e*, *e*, without these traverses no redoubt would hold long against heavy artillery.

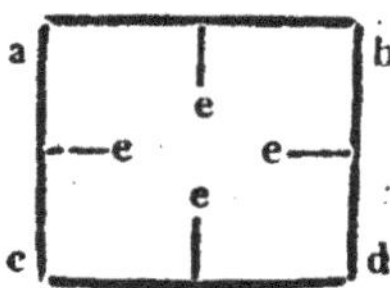

SECTION IV.

Of the Defence and Attack of single Field-works and of Entrenchments.

CHAPTER I.

Defence of single Field-works and of Entrenchments.

§. 210. *Defence of single Field-works.*

The guns in the work should never engage those of the enemy, but should be removed from the platform, during his cannonade, and should be used only when he has arrived within 600 paces of the work, when they should commence their fire with case shot. Not having fired before, the enemy is not concealed from them by the smoke, and the men are not fatigued; the effect will now therefore be very considerable. The distance of 600 paces should be previously marked with small stakes, that the men may not, as is frequently the case, open their fire too soon.

They should remain during the enemy's cannonade, at the foot of the banquette, which they should ascend when he has advanced within 250 paces of the work; this distance should therefore be likewise marked by small white stakes. The garrison should fire by divisions; each side of the work should be formed in 2 platoons, which should fire alternately*. If the ditch is pallisaded, it would perhaps be best not to commence firing until the enemy is within 50 paces; and when the work has trous-de-loup or abbatis, besides pallisades in front of the ditch, the musquetry certainly ought not to open until he has reached the first of these impediments. The great and sudden effect of the fire at this short distance, at the moment when his advance is checked, will, if there has been no firing before, operate powerfully upon the minds of the men. At a greater distance, the enemy occupying a greater extent, the fire from the fort spreads, and the slight effect which is produced by it, emboldens him, at the same time that the men cannot so much notice the cries and groans of their wounded comrades, as they immediately advance from the spot.

b) When the enemy enters the ditch, shells with the fuses lighted, or leather bags filled with 1 1/2 or 2 lbs. of powder, and provided with very small fuses, should be thrown into it; and at the same time, he should be received with bayonets, fixed to the end of long poles, when he is about to attempt the escalade. If not provided with these means, the men should mount the crest of the breastwork to receive the enemy with the bayonet; this, however, must not not be done

---

* If every man was to fire independently and without order, the ammunition would be expended soon, and with, perhaps, very little effect. The disadvantage of stepping down frnm the banquette to load, and the manner of firing, employed in some services, have been pointed out in my Manuel, Vol. II, section 217.

until the whole, or at least, the greater part of them, have entered the ditch*.

c) If there are a greater number of men than are requisite to man the work, when formed 2 deep, or if more can be procured, one, or, if possible, two ranks should be placed in the ditch behind the pallisades, with orders not to fire until the enemy arrives close to the pallisades. This fire, given at so short a distance, must prove most destructive; and together with the fire from the breastwork, its dreadful and unexpected effect will certainly compel the enemy to retreat.

d) It is necessary that the officer should expose himself, in order to keep up the courage of the men, he should be the first upon the breastwork, &c.; he should tell them that their preservation depends upon their gallantry, and that formerly forts were defended solely by the sword and spear or pike, and that they must ultimately succeed by resolution.

e) For security against a night attack, nothing can be done, but to keep a part of the garrison constantly under arms, near the breastwork, and to post a chain of sentries at the distance of at least 300 paces round the work; and further in front, at proper places, single men, for the purpose of listening, particularly in the rear. It is advantageous to have piles of wood prepared at 50 paces from the ditch, which may be set on fire, as soon as the enemy advances: for this purpose, however, it will be necessary that the wood should be short

* In order to ascertain that these orders are understood, you should cause the work to be attacked previously by a few men, who will represent the enemy; you fire upon them with blank cartridge, and they then attempt the escalade, etc. Without this, the orders alone will make no impression.

and dry, together with a quantity of dry straw, and that it should be covered with a small roof, or else it will probably not burn until the work is taken. Some persons have recommended, that hay and straw should be kept in readiness in the ditch, and to set fire to it when the enemy approaches. If the work is weak, has a small garrison, and may be quickly supported, this might be of use.

§, 211. *Defence of Entrenchments.*

a) The camp is from 300 to 600 paces in rear of the entrenchments. Each battalion has a certain extent allotted to it, which it will occupy, drawn up 2 deep if possible, as soon as any alarm takes place; and to which it furnishes a guard during the day, and a strong picquet at night, and in case of danger, supports it with half of the battalion.

b) The regimental guns are placed in the works with their battalions; the others are divided into batteries, which are distributed at intervals of from 800 to 1000 paces. In the entrenchment, Fig. 10, Pl. V. 8 guns are placed in the work, *a b*, 4 on each side; and in the entrenchment, Fig. 11, there are 4 guns in each of the redoubts in the first line, Nos. I, III, and IV.

c) A strong reserve of infantry and artillery must be stationed at some distance, in the rear of the works, to support the part attacked, or to repulse the enemy, should he penetrate any part of the lines; to this latter service, the cavalry is particularly destined, and should be for that purpose formed, in bodies of 8 squadrons each, at from 600 to 1000 paces behind the entrenchments, during the enemy's cannonade, and should attack the enemy the instant he has entered the lines.

d) The batteries may open their fire on the enemy's guns at 1200 paces, and some stakes or other marks should therefore be fixed at that distance; but they should on no account fire at a distance at which the enemy's fire does not take effect; and the same must be observed when the enemy is superior in artillery. It is well to conceal a few batteries, until the enemy's troops have approached within 600 paces; therefore the fire should at first be only opened from the batteries of least importance, in order to draw off the enemy's fire from the principal ones, that they may not suffer, and may then receive the enemy's troops unexpectedly with case shot. The ammunition should, if possible, be kept covered in small magazines, at 20 paces or more in rear of the batteries; or the ammunition waggons must be stationed further back, and separate from each other.

e) With respect to the remaining part of the defence, see §. 208, and for precautions of security against surprise, see §. 64, 65, 66, 67, &c.

CHAPTER II.

Attack of single Field-works and Entrenchments.

§. 212. *Attack of single Field-works.*

a) *They should be surprised*, by turning them, and then rushing in through the entrance, in this case the directions in §. 60 are to be followed.

b) *They should be attacked at night*, when you are

not provided with artillery, and the enemy can be succoured quickly. It is always best to make the attack at midnight, in as many divisions as the work has sides: in addition to which, if the enemy has any support, one division should proceed against it. The different divisions must try to enter the work undiscovered: but if they are discovered, they should advance as rapidly as possible, and endeavour to rush instantly into the work by the entrance and by the ditch: this in general succeeds, if the garrison is not extremely vigilant. Each division should, however, be accompanied by persons carrying planks and ladders, and by carpenters with axes, to remove the pallisades, &c. If you fire during the attack, all is lost; it is necessary, therefore, to make the men previously draw their charges.

c) *In a regular attack*, the work should first be cannonaded during the day; the guns and howitzers should be formed at 900 paces from the works, and commence firing *en ricochet*, so that the shot may reach the work at about the 3d or 4th graze. The howitzers may also be elevated to an angle of 20 degrees, and shells thrown from them into the works, the fuses of which should be so prepared that they may explode the instant they fall. In this case, however, they should approach within 700 paces, and if the work is not provided with heavy artillery, within 500 paces. Whenever opposed by the enemy's artillery, the guns and howitzers should be placed 30 paces from each other.

After cannonading the work for 2 or 3 hours with a certain proportion of ordnance, the artillery should approach, the guns being extended, to within 600 paces, and if the enemy has no heavy artillery, to within 400 paces of the work, and then commence firing as quick as possible with case shot; 20 or 30 rounds should be fired from each piece; immediately after which, the infantry, which until then has remained in rear of the flanks of the artillery, at 900 paces from the work, should advance to attack the work in as many divisions as it has sides, and in all directions at the same time without firing; the artillery continues its fire until the infantry has passed it, and then retires. Each division is formed in 3 subdivisions; the first is destined to the escalade; the 2d is to remove the impediments, and is provided with axes, ladders, spades and fascines, into which latter, wooden spars are bound to prevent their bending too much; the 3d division fires, while the 2 former pass the trous-de-loup, &c. leap into the dith and scale the work. If the first division were to fire, the assault might be interrupted. The work should be scaled at the angles upon a given signal; the officers should be foremost.

§. 213. *Attack of Entrenchments.*

a) The attack is ussually made at day-break, and the work approached during the night to the distance at which it is intended to commence the cannonade and bombardment. At night there would be less to fear from the enemy's fire, but on the other hand confusion would be more likely to take place.

b) The attack is made at several points; 1 or 2 battalions of grenadiers generally form the advance at each; then follow 20 or 30 carpenters with axes, 100 men with spades, and 200 with ladders and fascines; these are succeeded at some paces, by 4 battalions drawn up in line, and at 2 or 300 paces in rear of them, 6 battalions likewise in line, which are covered, towards the flanks, by several sqadrons of cavalry.

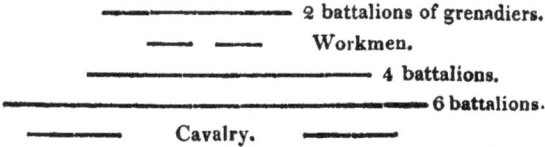

——————— 2 battalions of grenadiers.
—— —— Workmen.
——————————— 4 battalions.
————————————— 6 battalions.
——— Cavalry. ———

c) The cannonade and bombardment are to be carried on, according to §. 212. c, for several hours, and the attacking parties are then to advance without firing; the artillery is alone to keep up a constant fire of caseshot until it has approached within 400 paces of the work.

d) For the rest, the actual attack is conducted according to the foregoing paragraph; to which, however, is to be added, that a passage should be formed across the ditch with the fascines, the earth is to be cut away from the edge of the ditch and the breastwork, to enable the guns to pass. The workmen should be instructed in this service, and should be led by officers of the engineers, who are to direct the work. Others should cut away the earth to the right and left of the ditch, for the infantry to cross. The troops that pass first must form upon the summit of the breastwork, extending themselves to the right and left, and the battalions should then advance regularly with their guns.

e) Part of the entrenchment should be assailed on both sides by 2 or 3 of these attacking divisions in order to get the enemy under our fire in all directions, and that the troops, which penetrate first, may be able to join each other. The attack should also be made against those parts of the entrenchment, by gaining which we may become master of the whole; other wise, the troops are sacrificed without any thing being effected.

f) The army follows the attacking divisions in the usual order of battle.*

---

* The disposition of General Laudon, for the attack of the entrenched camp near Bunzelwitz, was in substance as follows: Two divisions were at day-break to attack, near to each other, an angle of the entrenchments, and these were to be supported by the whole army formed in 3 lines. One of the attacking divisions consisted of 50 volunteers, which formed the advance; then followed 400 volunteers who were to storm both sides of the work, which it was intended to attack; these were accompanied by 50 workmen with axes, nails to spike the guns with, etc. The 2 battalions of grenadiers were in line, after these, 4 batallions in line, and lastly, 2 batallions and 5 squadrons in line. The first line of the army consisted of 10 squadrons and 16 battalions; the second, at 400 paces in rear of the first, of 30 squadrons and 12 battalions, and the third of 30 squadrons and 4 battalions. The attack was to begin an hour before day-break, and the army was to support the attacking divisions in case of success. Thus no cannonade was to take place beforehand.

SECTION V.

Examples of the Situation of Field-works, of Entrenchments, of Fortifying a Building, Village, &c. and of the Attack and Defence of the same.

CHAPTER I.

Field-works.

§. 214. *Field-works near a Ford.*

It is intended to render the two fords Fig. 6, Pl. V. impracticable, and then to construct a work at *a*, for 50 men, from whence the enemy may be prevented from re-establishing the fords.

The distance of the enemy is 2 miles, and that of ou nearest corps about the same. There are bridges on the right which are occupied; and on the left there are no fords, nor any parties of the enemy for several miles.

Iron harrows (to which bags of stones are fastened) and cart, wheels, should be sunk in the fords; and a trench dug in front into which the water should be afterwards let. Strong stakes should be driven in the bottom of the ford, and trees, with the branches cut off, thrown in a little above, and fastened to these stakes. Waggons loaded with stones, should be brought into the ford, and some of the wheels taken off, or knocked to pieces.

The work for the defence of the ford, should be closed, otherwise a party of the enemy, having crossed the river in a boat, or on a raft, might surprize at night the small detachment posted in it. The breastwork should be 10 feet thick at top towards the river and 5 feet on the flanks: the ditch should be all round 10 feet wide at the top, and 9 feet deep, and should be filled with water. The sides of the work, when the angles are cut off, should be 18 paces in length; the angles towards the river should however be cut off 4 paces, and will then remain about 12 paces in length, between the points where the angles are cut off, so that 40 men, two deep, may fire at the fords from this side, and the parts where the angles are cut off, and yet be able to occupy the whole work when formed in single rank.

During the day 2 or 3 sentries may easily watch the enemy's approach; but at night half the detachment should be constantly under arms; and from 4 to 6 centinels should be posted round the work at 300 paces distant; the entrance should be closed and the plank over the ditch removed; sentries should be placed at the angles and in front of the entrance.

In the attack of a post of this kind, it is necessary to cross the river secretly, perhaps at night; after having procured the necessary information according to §. 43, to approach the work quietly in the night, and rush quickly into it, without firing. The most proper time for the attempt, is at the moment when it becomes dark in the evening, and the bridge over the ditch is probably not removed. Rainy and stormy weather will also favour a surprize.

§. 215. *Fieldwork near a Bridge.*

An outgard of 100 infantry is ordered to defend a bridge. (Fig. 9. Pl. V). The army is cantonned near the river, on the banks of which there are posts at 2 or 3000 paces to the right and left. The bridge is to be maintained as long as possible, that the light troops on the opposite bank may be drawn in.

A pile of dry wood with straw should be placed under the bridge, and one immediately behind it; so that if the latter is set fire to, no person will be able to pass; and if the former, the flames will communicate to the bridge itself. A fleche should be thrown up in rear of the bridge, behind a garden paling, that the enemy may not be able to dislodge the party; the pales should be pierced with holes according to the direction necessary for the fire. In the day, an officer with 20 men is stationed beyond the bridge in a small fleche *b*, who will examine every thing that approaches. A double sentry will be posted 200 paces in front, and another on the bridge, the planks of which should be loosened for about 12 feet, and the men directed to throw them into the river, in case it appears necessary; as the officer and his party may retire over the beams. This post will be drawn in during the night, and the planks, which have been loosened, removed from the bridge; 2 or 3 sentries should be placed near the river, and some at 400 paces round the post. If this detachment is only relieved every 3 or 6 days, two thirds of the party may remain during the day, and one third during the night, in the castle at *c*, which however should then be included within the chain of sentries.

The faces of the fleche behind the bridge are 30 paces in length; the earth is taken from within, as in Fig 5, Pl. VI., and thus little work is necessary.

This post cannot easily be taken, if properly defended. An attack will probably succeed in rain or in fog; having arrived undiscovered at the sentry, every endeavour should be made to reach the bridge at speed, which should be kept possession of by a part of the troops, while the post is attacked in the rear by the remainder; but enterprizes of this nature are generally mere attempts. If the possession of the bridge was of importance; the river should be passed at night in boats, and the post attacked in the rear; the bridge may then be opened to other troops, who should have arrived in front of it at a fixed time.

§. 216. *Of Field-works erected on Hills or Hillocks.*

There are three cases of this:

a) A breastwork is to be placed on the brow of a hill, so as to command the ascent. Here the situation and size of the work.

depend on the form of the hill: a proper situation may however be found, sometimes by placing the work lower down the side of the hill, and sometimes by bringing it nearer to the summit, so as to be able to fire in every direction. Almost all the works, from No. I. to XIII. inclusive, of the Colberg entrenchments, Pl. II. No. I. are situated on hills in this manner; so also was the redoubt in the famous camp at Bunzelwitz. If the work is connected with others, it is frequently of importance that the enemy should be under its fire at a certain and perhaps a great distance, in which case the work must be placed accordingly. A principal object, and never to be lost sight of, is, that the enemy should be under, at least a direct fire, while passing the trous-de-loup and when he arrives at the ditch; if this is not the case, the work cannot be expected to maintain itself; and for this, if nothing else can be done, the breastwork must perhaps be placed higher up the hill, if it be possible to find a situation from which it may afford the necessary defence to the adjoining works. If there are hollows on the slope of the hill, where the enemy cannot be seen, fougasses should be established there; and abbatis should be laid exactly at the point where the enemy comes under fire, on quitting the hollow, (this was the case with the entrenched camp near Bunzelwitz) he will then find great difficulty in forming, particularly if there are trous-de-loup in front of the ditch. The profile Fig. 4, Pl. VI. may often be made use of.

b) A work is to be placed on the summit of a hill, not on the brow, but towards the centre: by this it is to be understood that the summit of the hill is nearly flat. This situation is very favourable for single works: it cannot easily be cannonaded, and the enemy, on arriving at the brow of the hill comes immediately under the fire of the work, which, at so short a distance, must be very destructive, and he is also impeded by the obstacles prepared there.

A redoubt thus situated, with trous-de-loup in front, (a little below the brow of the hill,) and having a ditch with pallisades, is always very strong. There were several in the entrenched camp near Bunzelwitz, placed in this manner.

c) A breastwork is placed near the brow of a hill. This must often be the case, when the hill is narrow; and then nothing is to be done but to adopt the profile of Fig 4, Pl. VI. The interior space of works of this description is not secure, and for that reason, they are only made use of as batteries in extensive lines.

d) If a very steep hill is to be fortified; the breastwork should be placed on the steepest spots, which should be rendered still more so by cutting away part of the hill.

e) All these four cases may frequently be combined, as far as circumstances will admit, without however losing sight of a maxim, the result of experience: that any work is lost, which the enemy can reach, without being obliged to pass impediments close under its fire.

§. 217. *To fortify a House, Castle, Church, &c. and of the Defence and Attack of the same.*

a) *Fortification and Defence.* If the walls are weak and of the thickness of only one stone, they should be strengthened within to the height of 6 feet, with strong planks or wooden spars, 4 or 5 inches square. It is to be understood that these spars should be firmly proped up. All the windows and doors should be barricaded in the same manner. Loopholes should be made through the wall, in the lower story, not more than one foot above the ground, and in the upper story about 7 or 8 feet from the ground, as shown in Fig. 21, Pl. VI.; with respect to the former, the earth is excavated at *a*, to enable the troops to fire through them conveniently. The loopholes are 4 inches wide without and one foot within. Stones should be carried to the top of the building, where apertures should be made in the walls, through which they may be thrown upon the enemy, when he attempts to storm. The roof should be covered with earth or dung, and every thing inflamable should be removed. If the building is strong and capable of resisting heavy artillery, the roof may be taken off, and the rafters, beams, stones, &c. laid upon the upper floor, which should then be covered with earth or dung, to the thickness of several feet. Water should be kept ready every where in large vessels, but particular in the upper part. When a moat or a dry ditch of considerable depth and pallisaded, can be formed round the building, and there is a sufficient store of shells filled, with fusees, and of carcasses, its occupation may be considered as tolerably secure against any attack. The carcasses are thrown upon the enemy, when he tries to scale the building during the night, and the shells are flung into the ditch whenever he attempts a scalade.

The garrison is distributed in the different rooms: a reserve is stationed in the middle of the house for the purpose of supporting the part principally attacked, and a party is on the roof with the stones.

Provisions and ammunition are the first things to be attended to.

b) Attack If a building is to be attacked without heavy artillery, an attempt must be made to surprize it in the night, in order to prevent the enemy from deriving the advantage, he otherwise would do, from his fire. The troops should be formed in 6 or 8 divisions, and the building assailed on every side. Each division must be accompanied by some men with ladders, and others with axes, &c. to break openings in the walls, or, if the building is strong, the barricaded windows, into which grenades shouldbe thrown to create confusion, and when they burst, the assailants should rush in. Perhaps a sort of moveable penthouse may be made use of, under which the men may be sheltered from the stones above and the musquetry in front.

§. 218. *To put a Churchyard and a Town surrounded with a Wall in a state of Defence, and to attack the same.*

a) Fortification and Defence. The wall of the churchyard, if from 9 to 12 feet high, according to Fig. 23, Pl. VI, should be furnished with loopholes and a ditch. If it is only from 6 to 8 feet high, a ditch may be formed in front, and a banquet raised behind it, for the men to stand on and fire over the wall; the scaffolding *a b c*, which is intended for the same purpose, is then unnecessary. This scaffolding is formed by the post *a b* and the joist *c*, which is fastened to the post and fixed in the wall. Scaffoldings of this sort are erected from 8 to 8 feet distance from each other, and are connected by laths, nailed from post to post, and planks are then laid across the joists. In front of the entrance to the churchyard, a tambour is raised: this is a work formed with pallisades, as shewn in Fig. 22. The pallisades used for this purpose, are 10 feet long, and 4 1/2 inches thick; they should be fixed 3 or 4 feet deep in the earth, and close together; loopholes having been previously cut at 2 feet asunder, between every other two pallisades, and at 6 feet from the ground.

In Fig. 3. Pl. V. a section of a tambour of this kind may be seen at *a* and *c* and at *bb*, the elevation of the same, in which the height of the loopholes is marked by *bb*: the loopholes are 8 inches wide within, and 4 without. At *c* is a banquette, from which the troops may fire conveniently through the loopholes. A tambour is chiefly of use against an escalade, and it affords a flank defence.

But to render the wall of the churchyard completely secure against an escalade, it is indispenisible that the ditch should be pallisaded, and a small breastwork should also be erected behind the entrance, as represented at *a* Fig. 22. Pl. VI.

It is not easy to provide the wall of a town with a scaffolding in every part; instead of which therefore, planks should be laid on common tressels, Fig. 8. Pl. V. which can always be brought to the place, where they are required, (the part of the wall against which the enemy erects a battery). If the wall, as is commonly the case, exceeds 2 feet in thickness, it will be injured by piercing it with loopholes. When there are rondels in the wall, as at *t*, *m*, *p*, *s*, *r*, No. 1, Pl. V. loopholes may be made in them, if they have not, as usual, been previously formed; if they are already prepared and are too low, the earth must be excavated to a sufficient depth; and if they must be broken through the wall, they should be afterwards lined with planks and spars, that the regular form of the loophole may be obtained.

Half redoubts should, when practicable, be erected before the entrances or gates, as at *e*, *dc*, and *a*, No. 1, Pl. V. These should have a strong profile, and should be provided with pallisades and trous-de-loup; on them the defence of the town depends, and one should be erected on each side, or at each angle, for the purpose of commanding the whole wall. The other arrangements necessary for defence are stated in §. 124.

b) *The Attack.* The works in front of the gates, are to be attacked, like other field-works, in the night; the walls should be scaled by means of ladders, at the same time with the works; but it is also necessary that the rope ladders should be provided for the purpose of descending on the other side. The escalade takes place during the night, at several different points; at one of which perhaps none of the garrison may be prepared, or too few to make any resistance. If any party succeeds, it should immediately attack one of the works in the rear, by which means they may be able to open a gate to the remainder of the troops employed in the assault.

A night attack of the above description can only be made when the garrison is weak, and not provided with heavy artillery; otherwise batteries must be erected to silence the guns in the enemy's works, and to batter down part of the wall.

For this purpose batteries must be established at *x* and *y*, and at H, No. 1, Pl. V.; these batteries are not more than 4 or 500 paces from the town, and are thrown up in one night. The breastwork is raised 3 feet only, and the earth is excavated within to the depth of 3 feet, according to Fig. 5, Pl. V.

It is possible to get at the battery H, undiscovered, but not to those at *y* and *x*. On that account, ditches *x b*, *y b* and *b* F, have been made, in which it is possible, during the day, to pass from F to the batteries *x* and *y*, being covered against the fire from the town. These ditches are 3 feet deep, and the earth excavated from them is thrown towards the town; and they are so constructed that they cannot be enfiladed by the shot from thence. After a number of howitzers grenades having been thrown into the redoubts and into the town, from the battery *x*, part of the wall being destroyed from *y*, and the gate *c d*, having been cannonaded from the battery H, and the garrison in the redoubt having been harrassed by the fire; the attack is commenced by the troops: part proceed towards the redoubts *e* and *d c*, and part advance against the wall between *e m* and *m d*. In the mean time, the batteries continue their fire against the town, on the flanks of the troops who are attacking.

§. 219. *Têtes-de-Pont.*

Tetes-de-pont are erected to secure the retreat of troops across a river in presence of an enemy, or to enable them to pass to the side in his possession. These works must, 1st, cover the bridge against the enemy's artillery; 2d, must be sufficiently large to allow waggons passing, without any delay being caused; and 3d, must have such a figure, that the works, situated immediately on the banks of the river, may be able to defend those in front, and that the batteries on this side may command the whole. This is the case in the tete-de-pont, *a c*, Fig. 4, Pl. V. which may therefore serve as an example. The lengths of the lines are marked in paces on the figure. Such a work will be advantageously situated, where one flank is covered by natural obstacles, as in this figure; the right is covered by a rivulet, which is dammed up; further in advance, at *z*, and to the left, works are thrown up; when retreating, these works are furnished with cannon, which check the enemy's advance, and prevent his attacking the few troops which retire last. When these troops are to fall back to the main work, their retreat will be covered by the guns mounted there; these open works therefore should not be beyond the reach of caseshot from the main work. The tete-de-pont *a c* will then be defended by the batteries *b* and *e*.

The works shewn at *z*, serve to check the enemy's advance, until several of the troops have passed the river. But on such occasions these works are thrown up during the night, and the principal entrenchment *a c*, is then either not constructed or only just completed.

The defence of the principal entrenchment is similar to that of other fieldworks: it is only to be remarked, that when the army retreats, the ditches of the tete-de-pont, and particularly the entrance, should be filled with straw, shells, wood, carcasses, &c, which should be set fire to when the last troops pass the bridge, and the enemy presses forward to destroy the pontoons and cut off the rear detachment. Fougasses in front of the entrance under the barricades may be of great use, the batteries *b* and *e* playing at the same time with grape.

If a bridge of importance is to be covered by a corps, while the army moves forward, it must be surrounded with similar entrenchments to those *a c* Fig. 4, Pl. V. and if the strength of the corps exceeds what is necessary for occupying such a work, two strong redoubts should in addition be erected on one side, close to the river, and situated so as to assist in the defence of the other works. These redoubts must, like other field-works, be provided with pallisades and trous-de-loup; they are connected with each other by ditches. Higher up, two rafts are placed in the river, on which small wooden breastworks are constructed.

There is a guard constantly on these rafts and they are furnished with ropes, poles, &c. to prevent the approach of fire vessels,&c. and to secure the pontoon bridge against them. These rafts should be made fast by strong ropes to trees on both banks of the river. The rope is passed once round the tree, so that when the shock takes place, it may be slackened a little; it is then passed several times round and fastened.

If a stone bridge or one of any other description is to be put in a state of defence against detachments of the enemy; the best mode is by constructing half redoubts, as represented in Fig. 7. Pl. V.

Attack. The usual mode of attacking other redoubts may be employed here; but as the principal object is to damage or destroy the bridge, to that the attention must be first directed. This is not so easily accomplished by fire machines, as by very large trees and pieces of timber fastened together, and floated down the stream. The heavier these masses are, the greater effect they will have; and the destruction of the bridge will be more speedily completed, the greater the quantity of wood floated down. It would be more advantageous if inflamable materials, as wood, straw, &c. or carcases were placed upon them.

During the execution of this, the tete-de-pont should be attacked on both sides, the attack on one side should be serious, the other only a feint.

§. 220. *To put a Village in a state of Defence.*

In No. 2. Pl. V. is a village fortified by means of the guard houses A and B, the fleche C, and the stone building D, together with the pallisades between them, which may check the enemy until the garrison can get under arms. If a village is to be defended against an attack, the works A B and C, as shewn in the figure, should consist of half redoubts, closed in the rear with a tambour (§.218). The redoubts should have a pallisaded ditch; besides which a line of pallisades, placed with their points inclining outwards, should be drawn round the whole village, with a double row of trous-de-loup in front. The river however, and the ditch near C, will supply the place of pallisades on those sides. The house D, must be put in a state of defence according to §. 217; and a breastwork should be thrown up to command the ford.

The works and the fortified building must command the trous-de-loup. If it is necessary to render the village still stronger; breastworks may be errected at different places between the works and behind the pallisades. But in that case, two or three battalions will be required for the garrison.

§. 221. *Project for entrenching a Corps or an Army, and of the dispositions necessary for its Defence.*

a) *General project of the Work.* If a corps stationed in the vicinity of an army, (within a few thousand paces) is to be entrenched, it is only necessary to construct some works in front, and upon the flanks; the rear being covered by the army. In this disposition however, some slight works should be thrown up in the rear, to prevent the enemy from penetrating suddenly.

b) When an army is to be entrenched, the works must either surround the whole, like those at the Prussian camp at Bunzelwitz in 1762; or extend along one side only, when either of the flanks is covered by water, a large morass, &c. This cover to the flank must however extend some miles to the rear, so as to oblige any corps of the enemy, which may make a detour to avoid these obstacles, to separate itself so far from the army, that it may be attacked and beat before it can be supported.

c) An entrenchment, which does not cover every part of an army, or which may be turned is nearly useless. The enemy may easily accomplish the latter with part of his army, because we are obliged to keep our whole force within our lines. This was the defect of the entrenched camp at Crofdorf in 1759.

An army entirely surrounded with works, may however be invested in the same manner as a fortress; as was the case with the Saxon camp near Pirna, in 1756. This may be avoided in the following manner:—1st, intervals must be left between the works, at such places as the ground will admit of moving on a large front; 2d, the lines should be carried along ground, where natural obstacles will separate the enemy's force, if he endeavours to surround the camp; 3d, the entrenchment should be appuyed upon a fortress, or should at least, be at a short distance only from one; by which a free passage is obtained in that direction, or else the enemy is obliged to invest a very considerable circumference. The former was the case with the entrenched camp at Bunzelwitz, where the fortress of Schweidnitz, afforded the Prussians an open communication, although the camp was 1000 paces from it. If the enemy had surrounded the camp within that distance, he would have had the fortress in his rear; and had he invested the fortress at the same time with the camp, his force would not have been equal to the extent. The case in which the circumference is increased, by the camp being placed near a fortress, is exemplified in the entrenched camp at Colberg, Pl. II. Here the circumference is also increased by the morass between the works, without any disadvantage to the entrenched camp.

Situation of the Works.

The situation of the works depends upon the ground, which will be best explained by an example. In the Colberg entrenchment, the works from No. I to IX, may on the whole be considered as one connected line, similar to that in Fig. 10. Pl. V.; the single forts, I. and II. &c. in the Colberg lines, are the same as the works *a* and *b*, Fig. 10.; the intermediate works are not however arranged in the same manner, but are disposed according to the ground: therefore, when a connected entrenchment is to be formed, the method prescribed in §. 208, Fig. 10. Pl. V. should be strictly attended to, and the rules, therein laid down, departed from no further than the circumstances of the ground render necessary.

If the country was level, where the works Nos. X, XI, XII, and XIII, Pl. II, are situated, 2 rows of redoubts, as shewn in Pl. V. Fig. 11, would have been preferable; this however the ground did not admit of, and therefore to command the country with the cannon, the hills were merely surrounded with a breast-work.

At Nos. XX and XXXI, connected works, according to Fig. 10, Pl. V. were not practicable; as the salient parts would have fallen upon the morass.

If an army is to be entrenched with single works, the mode laid down in § 208, is to be followed; namely, the works are to be erected at 600 paces from each other, and a second line is to be constructed at some distance in the rear, which may fire through the intervals of the first.

The works are left open in the rear, when there are troops in line behind them, ready to act offensively; but they are closed when there are not sufficient troops to draw up in a line along the whole entrenchment.

For further explanation. see §. 211.

§. 222. *Of the Entrenchments at Colberg, and their Defence.*

When the Duke of Wurtemberg in 1761, was posted with 12000 Prussians, near Colberg, to cover that fortress, and was threatened with an attack by the Russians; he entrenched him-self, as shewn in Pl. II from No. I to XIII; Nos. XX, XXXI, and XVIII, formed the actual entrenchment, and Nos. XXVIII and XIX, were advanced detached works. The parapet of the enclosed works was 16 feet thick, and revetted with fascines; the ditches were pallisaded, and in front of them were 3 rows of trous-de-loup and fougassses. The works from IX to XIII were con-structed last, but were, like the others, furnished with pallisades, trous-de-loup, and 3 fougasses. Owing to the weakness of the corps our weeks were employed in the construction of these works, before they were completed. The entrenchment was so extensive, that not more than one half of it could be manned; for this reason, inclosed works, connected by a breastwork, were preferable; as even if the enemy, superior in numbers, should penetrate in some part, the other works could still maintain themselves, until the remaining troops should be able to repulse him; and should the enemy make a vigorous and regular attack at any point, by means of trenches and batteries, the connecting breastworks being already constructed, might be completely manned by troops brought from some other point. The advanced works, Nos. XIX and XXVIII, are intended to check the enemy for sometime by their fire, before he can reach the principal entrenchment, and they are well adapted to this purpose; but if they are taken, it equally encourages the enemy, and disheartens our troops. No. XIX was stormed and taken, the enemy having found it open in the rear; No. XXVIII was also taken, but retaken again. Advanced works of this kind should be very strong and closed in the rear; they should be furnished with pallisades and trous-de-loup, and there should be the means within the work for the men to shelter themselves from the shells and grenades; traverses at least are absolutely necessary. The troops employed for the defence of the Colberg entrenchment, were posted immediately behind the works. Each battalion was appointed to the defence of a particular work, to which it furnished the guard day and night; and on their maintaining these works, the honour of the corps depended. According to the disposition made by the Duke of Wurtemberg, the troops were, during the cannonade, either to stand or lay down at the foot of the banquette, which they were to mount only at the assault. The troops were to support each other, according to the judgement of the generals, and the reserve was to attack the enemy in flank, or act as circumstances might require. When the enemy entered the ditch, the troops were to monnt upon the breastwork, and fire down upon them.

After several fruitless attacks, the Russians began at last to open their trenches and to erect batteries at *t z i*, &c.

Had the Russians wished to penetrate the lines they should have first driven the Prussians, by a tremenduous fire of heavy artillery, from the works XII and XIII, and should then have proceeded to storm. But as these works were strengthened by pallisades, trous-de-loup and fougasses, and were besides furnished with traverses, this would not have been easily done; and even if they had succeeded, they would then have been taken in reverse by the works XI and XXX.

§. 223. *Entrenchment near Bunzelwitz and its Defence.*

The entrenchment formed a rectangle, four miles in length, and two miles in breadth The works were all detached, and stood at the distance of from 200 to 600 paces from each other; In some places were small single fleches, and in others the works followed the line of the hills The fronts of the largest were from 300 to 400 paces in length; and nearly all were open in the rear and mounted with cannon. The infantry were formed in line behind the works, with the battalion guns. The works, in front of which there was not either a morass or lake, had a pallisaded ditch, with trous-de-loup or an abbatis (or both) and fougasses. The trous-de-loup and abbatis were placed principally in front of the intervals between the works; so that openings of 100 paces only remained. The works mounted with cannon formed a line of batteries in order of battle The trous-de-loup and abbatis checked the enemy under their fire, if he attempted to penetrate at any point, while the battalions and squadrons formed in line, advanced to repulse him.

When open and detached works are mounted with cannon, and a line of troops posted behind them; the enemy, should he penetrate any where, may be driven back before he can form. When therefore there is a line of troops immediately in rear of the works, this disposition is very good. It also gives the power of concentrating the whole force against any one point, and it enables the troops to retire in good order; the part attacked may also be occupied by infantry, while the garrisons are withdrawn from the other works, and the troops held in readiness to be employed to any other purpose. These arrangements afford great advantages to skilful generals in manœuvring with well disciplined troops.

The villages and woods, to the distance of 2 or 3000 paces from Bunzelwitz, were occupied by light troops, or picquets from the infantry of the line; and between them and the villages cavalry outguards were posted. The enemy being immediately in front, and surrounding the entrenchments almost on all sides; the infantry, or at least a battalion of each regiment, remained under arms behind the works during the night. About 30 squadrons were stationed without the entrenchments, at the part where an attack was most apprehended, and where there was the greatest danger, and threatened the attacking party by their rapid and unconfined movement.

§. 224. *Defence of single Works.*

a) The work XIX. Pl. II. near the sea, is occupied by 400 men, with some artillery; in the work XVIII. is a battalion, and the other works are also garrisoned: the enemy is however in possession of the wood. How are these 400 men to maintain themselves in the work XIX.*? If the work, as is the case here, be open in the rear, it must be closed, if only with pallisades or an abbatis; or in case of necessity it may be barricaded. A chain of sentries must be posted in a circle on the outside of the work, at the distance of 300 or 400 paces, surrounding both the rear and flanks. At from 500 to 800 paces in front of these, trusty men always 2 together, are stationed behind trees and bushes for the purpose of listening, who will be relieved from time to time; they are not confined to any particular spot, and change their stations frequently. Sentries are posted in the ditch at four different places, and during the night, 50 men remain constaantly under arms within the works. If the work is not closed, these men are posted in the entrance. In situations of great danger, half or even the whole of the garrison must be under arms during the night.

For the remainder see § 210.

---

* This was precisely the situation of this work when it was taken by the Russians.

APPENDIX.

Measure, Rate of Marching and Distance of the Enemy.

Long Measure.

ENGLAND. The common measure is the yard, which contains 3 feet or 36 inches; the foot 12 inches, and the inch 12 lines. A mile is divided into 8 furlongs, 1760 yards, or 5280 feet, or 2112 common paces, the pace at 2½ feet. A furlong is divided into 40 poles, 110 fathoms, 132 paces, 220 yards, 660 feet, 880 spans, 2640 palms, or 7920 inches. A marine league is 3 miles, and 20 leagues, or 60 nautical or geographical miles, or 69½ statute miles, compose a degree.

FRANCE. *Late system.* The common lieue or *lieue terrestre* was 2288,33; the *lieue de poste* 2000; the *lieue marine* 2850,41; and the *lieu moyenne* 2565,37 toises. The toise was 6 pieds, and the pied, generally denominated pied de roi, 12 pouces, or 144 lignes. *Present system.* The common lieue is 0,444; the lieue marine 0,556, and the lieue moyenne 0,5 myriametre. The toise is 1,949 metre, the pied 3,248 decimetres, the pouce 2,706 centimetres, the ligne 2,256 millimetres. The myriametre is 2,25 common lieues.

GERMANY. The Rhinland ruthe is the measure commonly used in Germany and Holland, and in most of the Northern States, for all military purposes, it is divided into 12 feet. The Rhinland ruthe is sometimes divided into tenths, or decimal feet, and the pace is made equal to 2 decimal feet, or $\frac{2}{10}$ of a ruthe. The geographical mile, which contains 364420 English, 23707 ½ Rhenish, 22829 Parisian, or 25385 Hanoverian feet, or 9519 common paces. The stunde is commonly calculated at 6000 paces. 34 Rhenish feet are equal to 35 English feet.

. Russia. The archine, or Russian ell, is divided into 16 werschocks, and is equivalent to 28 inches English, wherefore 7 yards make 9 archines. The werste, or Russian mile, measures 500 sachines, 1500 archines, or 24000 werschocks, and contains 3500 English, or 3400 Rhenish feet. The sachine is 3 archines, 48 werschocks, and is equal to 7 English feet.

Austria. The klafter, or fathom, is 3 ells, or 6 feet of Vienna. The foot 12 inch, and the inch 10 lines. The pace, according to Vega, is equal to 2⅔ feet.

Spain. The *estado*, *braza*, *toesa*, or fathom, is 6 feet. The *pie Burgales*, or foot of Burgos, is divided into 12 pulgadas, or inches, and the pulgada into 12 lineas, or lines. The cuerda is 33 palmos de vara, or 99 palmos de Ribera. The *legua* is 8000 varas, or 24000 feet of Burgos. The *legua juridica* is 5000 varas, and is divided into 3 millas, or miles, the milla into 8 estadios, and the estadio into 125 pasos of 5 feet each. The legua maritima, or marine league, is 6666 ⅔ varas. 20 marine leagues, or 60 miles of Spain, make 1 degree.

Portugal. The vara consists of 5 palmos menores, and the covado of 3 palmos craveiros; the foot of Lisbon is half of the covado; 21 varas of Portugal make 34 covados of Lisbon; 5 varas of Portugal make 6 yards of England; 20 yards of England, 27 corvados of Lisbon.

Sweden. The mil, or mile, is reckoned 18000 Swedish ells; the ruthe is 8 ells, 16 feet, or 192 inches. The famn, or fathom, is 3 ells, or 6 feet; the aln, or ell, is 2 feet; the fot, or foot, is 12 tum, or 144 linear.

Denmark. The common mile is nearly equal to a geographical mile, measures 12000 ells, and contains about 22000 Parisian feet.

Prussia. The Rhinland measure is used. See Germany.

Saxony. The ruthe contains 15 feet 2 inches; the Saxon ell, which is used instead of the pace, contains 2 Saxon feet. The Saxon police mile is rather more than 1¼ geographical mile, or 28341 Parisian or 29333 Rhenish feet.

Hanover. The Callenberg ruthe is generally used It contains 16 feet, or 6 paces, each pace equal to 2 feet 8 inches. A mile contains 12000 paces, or 2000 Callenberg ruthen or 32000 Calenberg feet. The foot is divided into 12 inches, and the inch into 12 lines. The same equation of measures is used in Brunswick.

TABLE

OF THE

LENGTHS OF THE MEASURES

OF

DIFFERENT NATIONS

IN ENGLISH FEET.

Places.	Measures.	Length of each Measure — Feet.	Equivalent to 1 Degree — Num. 100.
Denmark . .	mil	24704	14.75
England . .	mile by land . . .	5280	69,02
	mile by sea . . .	6073⅔	60
	league marine . . .	18221	20
France . . .	lieue terrestre or common	14576⅕	25
	lieue moyenne . . .	16398	22,22
	lieue de poste . . .	12784	28,50
	lieue marine . . .	18[illegible]21	20
	myriametre . . .	32797	11,11
Germany . .	meile	20587	17,70
	meile geographical .	24294⅓	15
Holland . . .	meile	19212	18,97
Hungary . .	meile	27378	13,31
India	parasang	12147⅓	30
Ireland . . .	mile	9110½	40
Italy	milla	6073⅔	60
Persia	parasang	16356	22,28
Poland . . .	meile	18221	20
Portugal . .	legua	20245 5/9	18
Prussia . . .	meile	25409	14,34
Russia	werste	3500	104,12
Scotland . .	mile	5952	61,23
Spain	legua of Castille . :	21958	16,60
	legua juridica . . .	13724	26,55
	legua maritima . .	18221	20
	milla maritima . .	6073⅔	60
Sweden . . .	mil	3[illegible]050	10,40
Switzerland	meile	27450	13,28
Turkey . . .	berri	5476	66,55

Rate of Marching.

INFANTRY.

1) A small party may march for a short time, that is, or 1 or 1¼ hours, at the rate of 100 common paces in a minute; and can, consequently, perform 9000 paces, (nearly a common German mile, or 4 English miles) in 1½ hours; but it is not possible to march more than 5 times that distance in 10 hours.

2) Large detachments cannot march even at the above rate; and it is frequently impossible for an army, in an intersected country, to perfom more than 2 common German (8 English) miles in a day. The rate of march of an army, is generally from 2½ to 3 German (10 to 12 English) miles in a day; in particular cases however 4 or even 5 German (16 or 20 English) miles may be performed in a day, when there are no defiles to pass, and when the army moves in several columns.

CAVALRY.

1) A small detachment may gallop for 5 minutes without halting, if the horses are in good order. Thus, calculating 500 paces to each minute; 2500 paces can be performed in 5 minutes.

2) When the above distance is to be exceeded, it is necessary to move at a trot, which however cannot be continued above ½ an hour. A party of cavalry may perform 300 paces in 1 minute, 6000 in 20, and 9000 in 30 minutes, at a trot. Thus a common German (4 English) mile may be performed in ½, or at the utmost, ¾ of an hour, provided the party is to halt immediately afterwards.

3) When a greater distance is to be performed, it is impossible, even when the roads are good, to march more than 6 German (24 English) miles in 12 hours, without fatiguing the horses too much; and when obliged to march 20 German (80 English) miles or further, not more han 8 (32 English) miles can be calculated upon in 24 hours; all the horses however will not be equal to this.

Distance of the Enemy.

1) *To ascertain it with a Telescope*.*

If a scale is fixed on the glass of a telescope, objects will appear after the telescope is adjusted, larger on the scale in proportion as they are nearer, and by making experiments, and marking the size of a man, that is the number of divisions a man covers on the scale, at different distances, you will in some measure be enabled, by means of the scale, to judge the distance of the enemy on other occasion.

2) *To ascertain the Distance of the Enemy with the naked Eye.*

At the distance of 2000 paces nothing can be discerned of infantry but the glittering of their arms; thus, if the colours, uniforms, files &c. are not distinguished, it may be considered to be at that distance. The files of cavalry will be seen at the above distance, but without being able clearly to perceive whether they are men on horseback. The files of infantry cannot be distinguished at a

* You will often be at a great loss without a telescope, and will be deprived of the advantage of making observations at a distance, as there is frequently an opportunity of doing in the field.

greater distance than 1500 paces, and at the same distance, the horses of the cavalry cannot be distinctly seen, but it may be ascertained that the men are on horseback. At 1000 paces the head may now and then be distinguished from the body, but it cannot be perfectly seen beyond 600 paces. The faces of the men and the lace and facings of their uniforms may be seen clearly at 3 or 400 paces, and at 70 or 100 paces the eyes of some of the men will appear like dots*.

*To ascertain the Distance of an inaccessible Object with accuracy***.

It is required to find the Distance of the inaccessible Object A from the Station C. Plate VI. Fig. 25.

First method without an instrument. 1) On the prolongation of the line A C, measure 50 paces from C to B, and at each of these points, place a picket. 2) Upon B C, raise a perpendicular by the eye, at C, make C E = 40 paces, and fix a picket at E. 3) On the prolongation of the line E A, make E F = C B (50 paces) and fix a picket at F. 4) Measure the line B F, deduct from it the length of the line C E (40 paces) and mark the

* From these observations, and what is said in another place respecting ranges it appears. 1) That caseshot may be used with effect, as soon as the head can be clearly distinguished from the body. 2) The fire of infantry will not be efficacious, until the faces of the enemy can be plainly discerned; and that in some cases (when you wish not to fire at a greater distance than 50 paces) you should not fire at cavalry until you can perceive the eyes of the soldiers.

** In reconnoitring any place or post, when stationed for some time in front of the enemy, or when it is intended to cannonade or bombard a place, in these and many other cases, it may be very advantageous to be able to ascertain with accuracy the distance of an object from a given station.

point D. If B F is found to be 42 paces, the difference will be 2 paces, and B D being equal to C E, D F will be equal 2 paces, and D E will be equal or nearly equal to B C. 5) Lastly C A will then be found by proportion, D F : D E : : C E : C A or as 2 : 50 : 40 : C A, which will be found to be 1000 paces*.

Second method with an instrument. To ascertain the length A C. Fig. 26, Pl. VI. measure a base A B (400 paces) and take the angle C A B = 40 degrees, and the angle C B A 64° degrees 10 minutes, then deducting the sum of these two angles from 180, the remainder will give the angle A C B = 75 degrees 50 minutes.

The sine of A C B : sine of C B A : : A B : A C - That is sine of 75° 50' : sine of 64° 10' : : 400 : A C. Then taking the sines from any table of lines, you will have 9,98658 : 9,95427 : : 400 : A C.

If the base A B is not less than ⅓ of A C, you will be able with a common theodolite to ascertain the distance with as much accuracy as is required here. The longer the base is, and the more accurately it is measured, as well as the angles, the nearer to the truth will be the result.

---

* In the practical application of this problem, the following circumstances must be attended to, 1) You should have a scale of one foot divided into inches, previously marked on your cane or sword. 2) The lengths should be measured with a staff, a lath, a cord or any thing that can most readily be procured; the length of the staff or cord should be measured by the above mentioned scale, as accuracy is here of consequence. 3) It is of great importance to proceed accurately, particularly in measuring the lines C E and B F; in this there should not be a mistake of a single inch, the picket which mark the Points E and C should not be too thick, and should be fixed perpendicularly. 4) If B C is made 120 paces, the greatest error that can arise in the result, will be 6 paces, when the object is 500, 12 paces when it is 1000, and 24 when it is 2000 paces distant.

TABLE

Of Corn Measure in English Cubic Inches.

Places.	Measures.	Contents of each Measure. Cub. Inches.	Equivalent to 10 Quarters Num. 100.
Alexandria . . .	rebebe	9578	18,19
Amsterdam . . .	scheepel	1647	105,77
Berlin	scheffel.	3315	52,26
Bremen	scheffel.	4336	40,19
Brunswick . . .	scheffel.	18963	9,19
Cassel	viertel	8702	20,02
Constantinople.	kisloz	2140	81,40
Corunna	ferrado	986	176,71
Denmark	toende	8481	20,54
Dresden.	scheffel.	6455	27,
England.	quarter.	17424	10,
	bushel	2178	80,
Florence	stajo.	1444	120,67
France	boisseau of Paris. .	774	225,13
	setier	9288	18,76
	décalitre*	610	285,64
FrancfortMaine.	malter	6584	26,46
Hamburgh . . .	scheffel. . . . :	6424	27,12
Hanover.	himten	1896	91,89
Leipsic	scheffel.	8473	20,56
Lisbon.	moyo	49440	3,52
	alquier	824	211,46
Malta - - - -	salma - - - -	16240	10,73
Munich - - -	schaff - - - - -	22109	7,88
Naples - - -	tomolo - - - -	3182	54,76
Nuremberg - -	simmern - - - -	20287	8,59
Persia - - -	attaba - - - - -	3974	43,85
Poland - - -	last - - - - - -	187260	,93
Rome - - - -	rubbo - - - - -	16684	10,44
Rotterdam - -	hoed - - - - -	67755	2,57
Russia - - -	chetwer - - - -	11888	14, 6
	chetwerick - - -	1486	117,25
Sardinea - - -	starello - - - -	2988	58,31
Scotland - - -	firlot wheat measure	2197	79,31
	firlot barley measure	3207	54,33

* The litre, or the unit of french measures of capacity, istherefore equivalent to 61 English cubic inches.

Places.	Measures.	Contents of each Measure. Cub. Inches.	Equivalent to 10 Quarters. Num. 100.
Sicily - -	salma grossa - - -	20215	8,62
	salma generale - -	16229	10,74
Spain - -	fanega - - - - -	3311	52,67
	celemine - - - -	276	631,30
Sweden - - -	tunna - - - - -	8932	19,51
	kanna - - - - -	159½	1092,42
Vienna - - -	metzen - - - -	4277	40,74
Wirtemberg -	scheffel - - - -	3228	53,98
Zuric - - - -	mutte - - - - -	5043	34,55

England. A last of corn measures 1¼ wey, 10 quarters, 20 cooms, 40 strikes, 80 bushels, 320 pecks, 640 gallons, 1280 pottles, 2560 quarts, or 5120 pints. The quarter consists of 8 bushels. A stone of hay, iron, shot &c. and horseman's weight is 14 lb. A tun of wine, of brandy and of other liquids, contains 2 pipes, 4 hogsheads, 6 tierces, 8 barrels, 252 gallons, 504 pottles, 1008 quarts, or 2016 pints. A gallon measures 231 cubic inches. A beer gallon measures 282 cubic inches.

France.---*Late system.* The pound consisted of 2 marcs, 16 ounces, 128 gros, or 9216 grains, the *tonneau de mer* was 2000 lb. and the quintal 100. The muid, corn measure, is divided into 12 setiers, and the setier into 2 mines, 4 minots, 12 boisseaux or 192 litrons. A setier of oats consists of 24 boisseaux or 384 litrons. The boisseau measures 16 litrons. A muid of wine is 36 setiers or 288 pintes. The setier is divided into 4 quarts, 8 pintes and 16 chopines.

Austria. The pound is divided into 4 vierting. A muth of corn consists of 30 metzen, 120 viertels or 240 achtels. The metze contains 1, 9471 Vienna cubit inches. The centner is 100 lb. and the stein 20 lb.

Russia. The pound is divided into 96 solotnicks, and the solotnick into 2, 4, or 8 parts; the czetwer, dry measure, is composed of 2 osmins, 4 pajacks, 8 chetweriks, or 64 garnitzen, 16 chetwerts are reckoned 1 last; and the chetwert is about 5½ Winchester bushels.

SPAIN. The libra or pound, is divided into 2 marcs of Castille 16 ounces or 9216 grains. The quintal consists of 4 arrobas, or 100 lb. of Castille, and the arroba of 25 lb. The cahiz, corn measure, contains 12 fanegas, 48 quartillas, or 144 celemines. The fanega is 12 celemines, and measures 43228 cub. inches of Burgos. The celemine is divided into 2, 4, 8, 16, 32 or 64 parts.

GERMANY. The pfund, or pound, consists of 16 unzen of 32 loths, 128 quintins 512 pfennings, or 1024 hellers. The dry measure is the malter, metze, scheffel &c. which vary according to the different states. The schipfund consists of 20 lispfund, and the lispfund of 14 lb.

PRUSSIA. The pound consists of 2 marcs, 32 loths, and each loth 4 quintleins. The centner is 110 lb. A last of corn is 8 wispels, and that of oats and barley only 2. The wispel measures, 2 malters, 24 scheffels; the scheffel is 4 viertels, 16 metzen, or 64 masgens.

SWEDEN. The pund or pound, is divided into 32 lods, the lod into 4 quintins. The centner is 120 lb. and the lispund 20. The tunna, corn measure, is composed of 2 spans, 8 fjerdings, 32 rappas, 56 kannar, 112 stops or 448 quarters.

DENMARK The pund, or pound, consists of 32 lods. The skipund consist of $3\frac{1}{5}$ centners, 20 lispunds or 320 lb. The centner is 100 lb. and the lispund 16 lb. A last of corn is 12 toende. The toende measures $4\frac{1}{2}$ cubit feet of Denmark.

HANOVER. The pfund consists of 2 marcs, 16 unzen, 32 loths or 128 quintlins. The centner is 112 lb. The schiffspfund is 20 liespfund or 280 lb. and the liespfund 14 lb. A last of corn measbres 2 wispels, 16 malters, or 96 himten of Brunswick, 6 himten make 1 malter, and 12 malter 1 fuder. A ration of oats is reckoned at $\frac{1}{3}$ of a himten (8 lb.) or $\frac{1}{4}$ himten of barley or $\frac{1}{2}$ of rye In Silesia 1 malter contains 12 scheffel

PORTUGAL. The pound consists of 2 marks, 16 ounces, or 96 outavas. The quintal is composed of 4 arrobas, or 128 lb. and the arroba of 32 lb. A moyo of corn consists of 15 fanegas, 60 alqueires, 480 selemis, or 960 mequias. The fanega is 4 alqueires, and the alqueir is divided into halves and quarters.

TABLE

OF THE

Principal Gold Coins of different foreign Nations, with their Weight and Value in English Money.

		Weight. Grs. 100.	Value. s.	d. 100
Bavaria, - - -	the carl - - - -	150,32	20	8,87
	the max - - - -	100,21	13	9,54
Bengal, - - - -	the gold mohur - -	176,50	30	10,95
Brunswick, - -	the carl - - - -	102,36	16	5,02
Denmark, - -	the ducat of 12 marcs	48,21	7	6,30
England, - - -	the guinea - - -	129,44	21	
	the half guinea - -	64,72	10	6,
	the 7*s.* piece - -	43,13	7	
Flanders - - -	the double souverain	171,50	27	9,79
	the souverain - -	85,75	13	10,89
France - - - -	the louis of 1726 -	122,90	19	7,65
	the louis of 1785 -	117,83	18	9,93
	the 40 franc piece -	199,25	31	8,85
	the 20 franc piece -	99,62	15	10,42
Geneva, - - -	the pistole of 1752	87,13	14	1,63
Genoa, - - - -	the zecchino - -	53,80	9	5,67
Germany, - -	the ducat - - -	53,85	9	4,78
Hamburgh, -	the ducat - - -	53,85	9	4,
Hanover, - - -	the georges - - -	103,03	16	6,31
	the gold gulden -	50,06	7	0,54
Holland, - - -	the ryder - - -	153,54	24	10,92
	the ducat - - -	53,85	9	4,
Hungary, - -	the ducat of Kremnitz - - - -	53,85	9	5,18
Madras, - - -	the star pagoda - -	52,75	7	7,03
Naples, - - - -	the onza - - - -	68,10	10	6,56
Piedmont - - -	the zecchino - - -	54,	9	6,09
	the pistole of 1741	110,10	17	7,92
	the doppia of 1755	148,50	23	9,83

GOLD COINS.

		Weight.		Value.	
		Grs.	100.	s.	d. 100
Portugal, - - -	the joanese - -	221,87		36	
	the moidore - - -	166,		26	9,35
Prussia, - - - -	the frederick - -	103,03		16	6,31
Rome, - - - -	the zecchino - - -	53,55		9	3,36
Russia, - - - -	the imperial of 1755	255,53		41	5,49
	the imperial of 1763	202,18		32	9,62
	the imperial of 1801	202,18		35	2,70
Saxony, - - -	the august - - -	102,		16	3,57
Siam, - - - -	the tical - - - -	281,88		39	8,04
Sicily, - - - -	the onza - - - -	67,94		10	10,77
Spain, - - - -	the doubloon before 1772 - - - -	416,65		67	4,87
	the doubloon of 1772	416,65		66	6,38
	the doubloon of 1785	416,65		66	0,74
Sweden, - - -	the adolphus - -	102,95		11	7,70
Tuscany, - - -	the ruopono - - -	161,33		28	5,45
United States,	the eagle - - - -	268,66		43	7,05
Venice, - - - -	the zecchino - - -	54,		9	6,09
Wirtemberg, -	the carl - - - -	150,32		20	8,87

TABLE

OF THE

Principal Silver Coins of foreign Nations, with their Weight and Value in English Money.

		Weight. Grs. 100.	Value. d. 100.
Aix la Chapelle,	the rathspræsentger - - - - -	95,68	7,85
Arabia, - - -	-the larin - - - -	74,17	7,93
Basil, - - - - -	the reichsthaler - -	436,89	53,38
Bengal, - - -	-the sicca rupee - -	179,55	24,92
Bern, - - - -	-the patagon - - -	417,63	48,59
Bombay, - - -	-the rupee - - -	178,31	24,38
Denmark, - -	-the ridsdahler - -	449,87	54,97
	the krohn - - -	344,	32,23
England, - - -	-the crown - - -	464,52	60,00
	the shilling - - -	92,90	12,
Flanders, - - -	-the ducaton - - -	513,29	62,34
	the croon - - -	456,91	55,26
	the patagon - - -	433,	52,91
France, - - -	-the ecu of 1726 -	452,50	57,40
	the 5 franc piece -	386,14	48,53
Geneva, - - -	-the patagon - - -	416,87	48,51
Genoa, - - - -	-the genovina - -	593,10	79,03
	the St. Gianbatista	321,66	41,17
	the giorgiuo - -	91,25	10,97
	the double madonina	140,19	16,45
Germany, - -	-the reichsthaler constitution money -	450,97	55,98
	the gulden ditto -	225,48	27,99
	the reichsthaler convention money -	432,93	50,38
	the gulden ditto -	216,46	25,19
	the old zweydrittel		31,98
	the new zweydrittel		27,98
Hamburgh, - -	-the rixdollar banco	450,52	55,92
	the marc banco - -	150,17	18,64
	the rixdollar lubs -	424,41	44,43
	the marc lubs - -	141,47	14,81

SILVER COINS.

		Weight. Grs. 100.	Value. d. 100.
Holland, - - - -	the ducatoon - -	503,50	65,91
	the three florin piece	488,	62,46
	the rixdaler - - -	433,17	52,93
	the leeuwendaler -	422,	43,70
	the gold florin - -	307,	26,26
	the current florin -	162,70	20,73
Madras, - - - -	the rupee - - -	178,88	24,61
Milan, - - - - -	the philip - - -	430,21	57,15
Naples, - - - -	the ducat - - - -	336,	42,81
Piedmont, - - -	the ducatoon - -	491,03	65,23
	the scudo of 1733 -	459,88	58,64
	the scudo of 1755 -	542,95	68,71
Pondicherry, -	the rupee - - -	177,27	23,83
Poland, - - - -	the tympfe - - -	89,75	6,44
Portugal, - - -	the cruzade - - -	265,65	33,31
Prussia, - - - -	the current rixdollar	343,42	35,97
Rome, - - - - -	the scudo moneta -	408,70	52,31
	the testono - - -	130,54	16,71
	the papeta - - -	81,59	10,44
Russia, - - - - -	the ruble of 1755 -	402,76	44,52
	the ruble of 1763 -	369,88	38,74
	the ruble of 1801 -	77,48	33,58
	the livonina of 1757	441,66	43,41
	the rixdollar albertus	433,17	52,93
Saxony, - - - -	the old reichsthaler	450,97	55,98
	the new reichsthaler	432,93	50,38
	the zweydrittelstücke	212,14	27,98
Spain, - - - - -	the hard dollar before 1772 - - - -	416,40	52,90
	the hard dollar since 1772 - - - -	416,40	52,09
Sweden, - - - -	the riksdaler of 1764	451,56	55,39
	the ducatoon - -	184,	62,30
	the carolin - - -	160,51	15,56
	the ten oere silver piece	108,30	6,72
Tuscany, - - - -	the francescono -	422,75	54,11
	the lanternina - -	420,	54,
	the livornina - -	402,	51,69
United States,	the dollar - - -	409,79	52,45
Venice, - - - -	the ducat - - -	350,83	40,42
	the scudo - - -	489,54	62,66
	the giustina - - -	433,17	55,45

TABLE

OF THE

Length, Weight, Diameter of Shot, Windage, Charge and Ranges of English Guns.

Nature.		Length in			Weight.			Diameter of Shot.	Windage.	Charge.		Ranges of the first graze of the Shot.			
		Calibers.	Ft.	In.	ct.	qrs.	lbs.	Inches.		lb.	oz.	P. B. yards.	1° yards.	2° yards.	3° yards.
24 Prs	Medium . .	16.483	8	0	41	3	2	5.475	.27	8	0	488	757	1103	1425
	Light . . .	13.000	6	3	24	0	—	—	—	3	0	162	364	606	722
12 Prs	Medium . .	16.872	6	6	18	—	—	4.403	.22	4	0		705	973	1189
	Light . . .	13.000	5	0	12	—	—	—	—	3	0		601	816	1063
6 Prs	Medium . .	18.500	6	0	8	3	27	3.498	·17	2	0		775	1003	1444
	Reduced . .	17.000	5	6	8	0	22	—	—	2	0		642	976	1150
3 Prs	Gl. Desaguliers	24.717	6	0	6	—	—	2.775	·14	1	0		679	883	918
	Light common	14.418	3	6	2	2	27								

TABLE

OF THE

Length, Weight, &c. of English Howitzers.

Nature.	Length.		Weight.			Length of Bore.	Chamber				
							Length.	Diameter		Powder contained in.	
								at top.	bottom.		
Inches. diam.	Ft.	Inches.	Cwt.	qrs.	lb.	Inches.	Inch.	Inch.	Inch.	lb.	oz.
10	3	11½	25	3	14	29.9	12.6	5.776	4.12	7	0
8	3	1	12	3	12	24.7	8.61	4.6	3.40	3	8
5½ Heavy.			10	0	0					3	0
5½ Light.	2	2¾	4	0	2	18.47	6.02	3.2	2.45	1	0
4 2/5 . .	1	10	3	0	13	15.21	4.52	2.7	2.24	0	8

RANGES

WITH A

Light 5½-Inch Howitzer.

Elevation.	Eight Ounces.		12 Ounces.		One Pound.	
	Range to first graze.	Extreme Range.	Range to first graze.	Extreme Range.	Range to first graze.	Extreme Range.
Deg.	Yards.		Yards.		Yards.	
P. B.	96	From 700 to 1000 yards.	140	From 1000 to 1350 yards.	159	From 1100 to 1400 yards.
1	143		334		325	
5	376		509		918	
8	620		975		1044	
11	797		1177		1173	

ERRATA.

Page 208, line 9 from bottom, read ***the mile equal* 2112 *paces.***

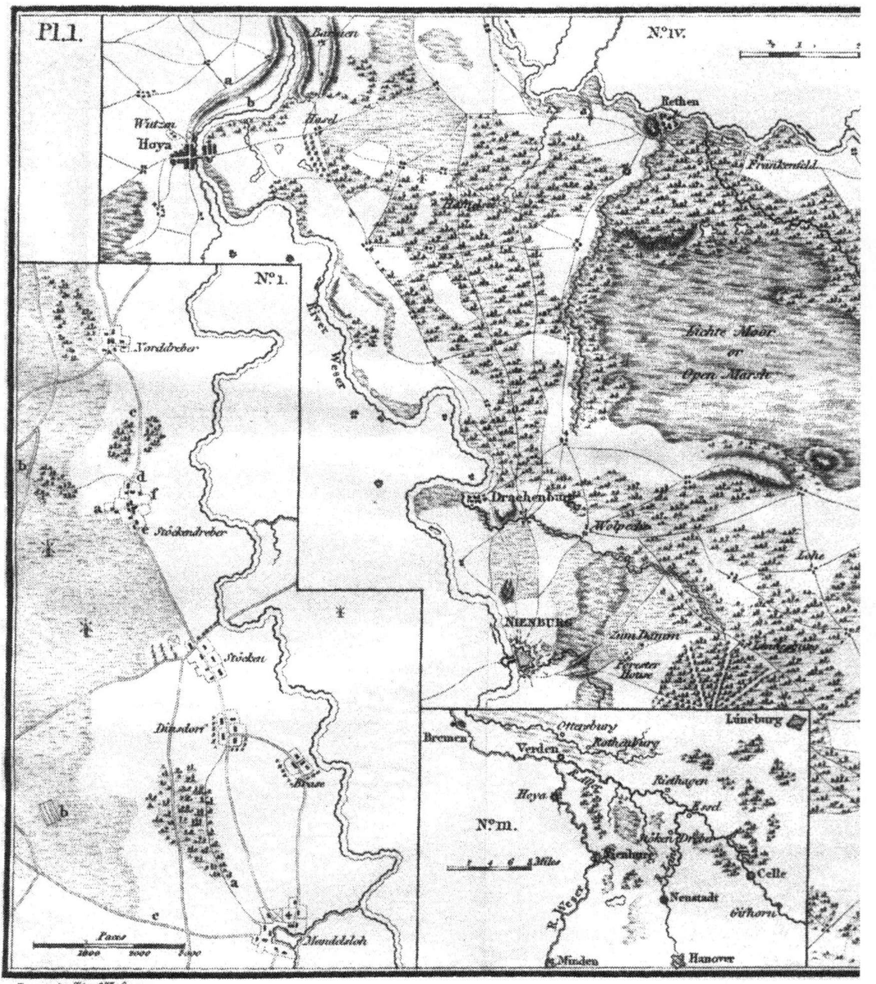

Drawn by Lieu.t Hofman

Published August 30th 1811 by T.E

Neele sc. Strand.

gerton, Military Library Whitehall.

Drawn by Lieut. Hofmann

Published August 30th 1811 by T. Egerton.

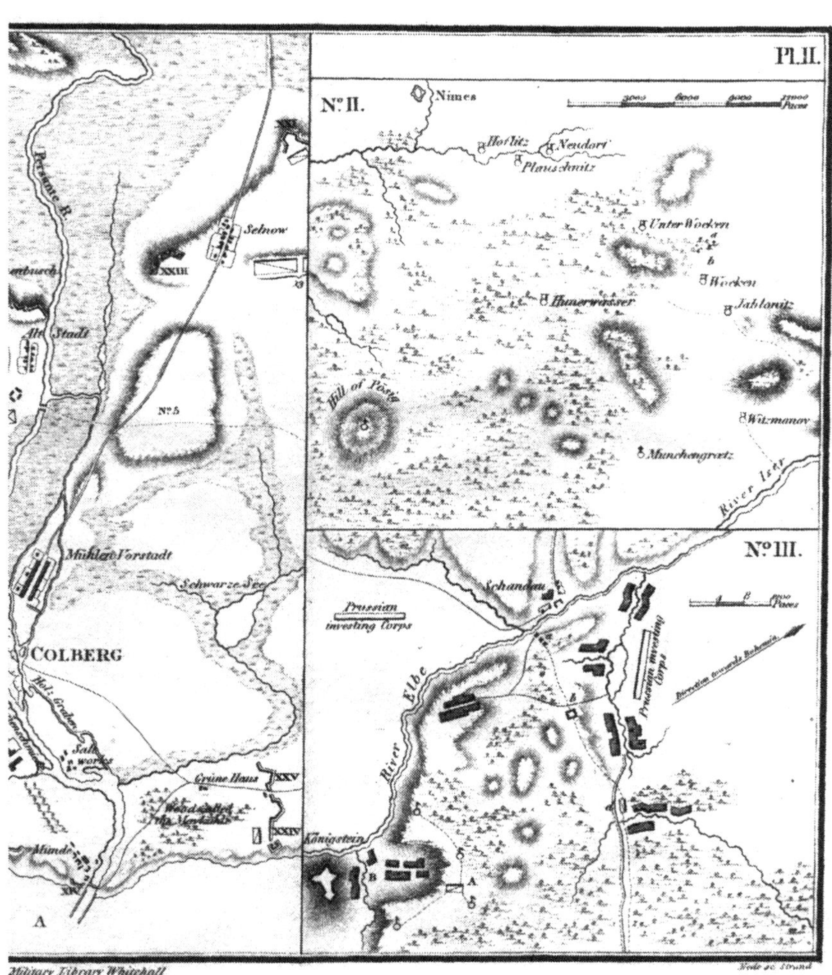
Pl.II.
Nº II.
Nimes
Hoflitz
Neudorf
Plauschnitz
Unter Wocken
Wocken
Hunerwasser
Jablonitz
Hill of Pösig
Witzmanov
Munchengraetz
River Iser
Nº III.
Schandau
Prussian investing Corps
River Elbe
Königstein
Direction towards Bohemia
Selnow
XXIII
Alt Stadt
Nº 5
Mühlen Vorstadt
Schwarze See
COLBERG
Grüne Haus
XXV
XXIV
Münde
A
Military Library Whitehall

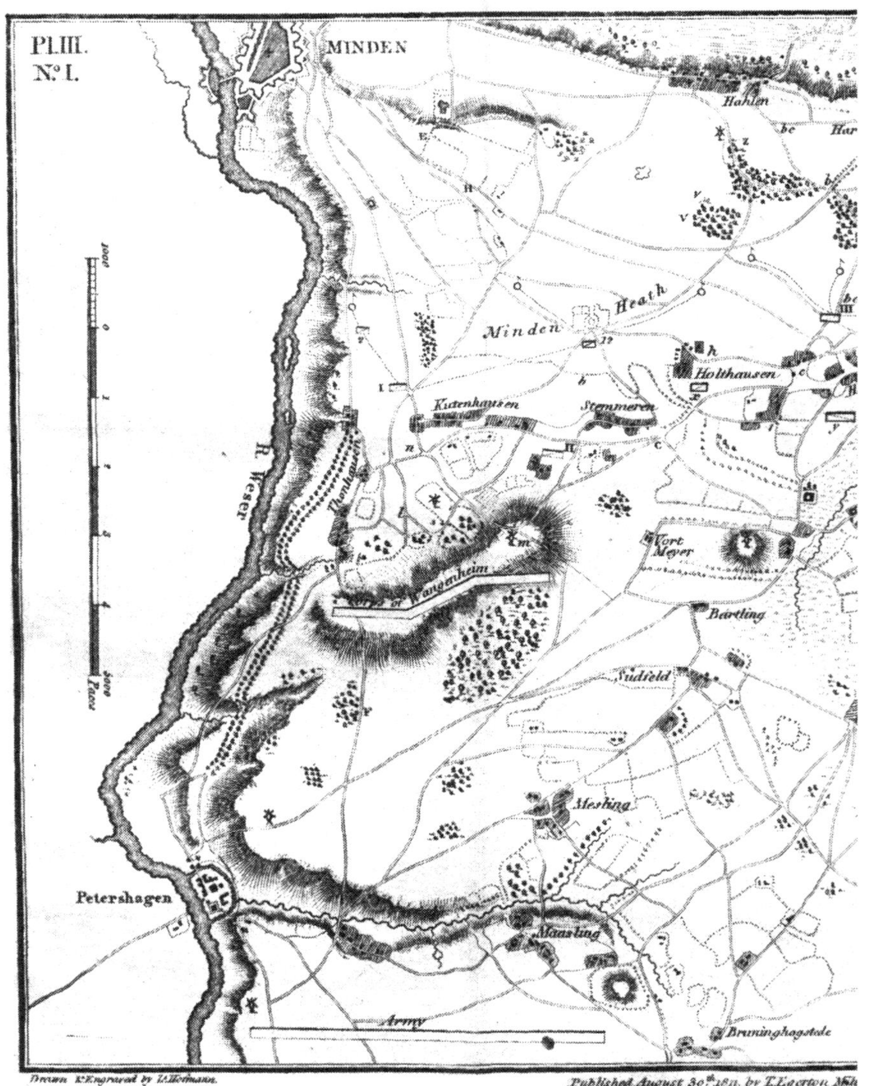

Drawn & Engraved by I. Hofmann.

Published August 30th 1811. by T. Egerton Mili

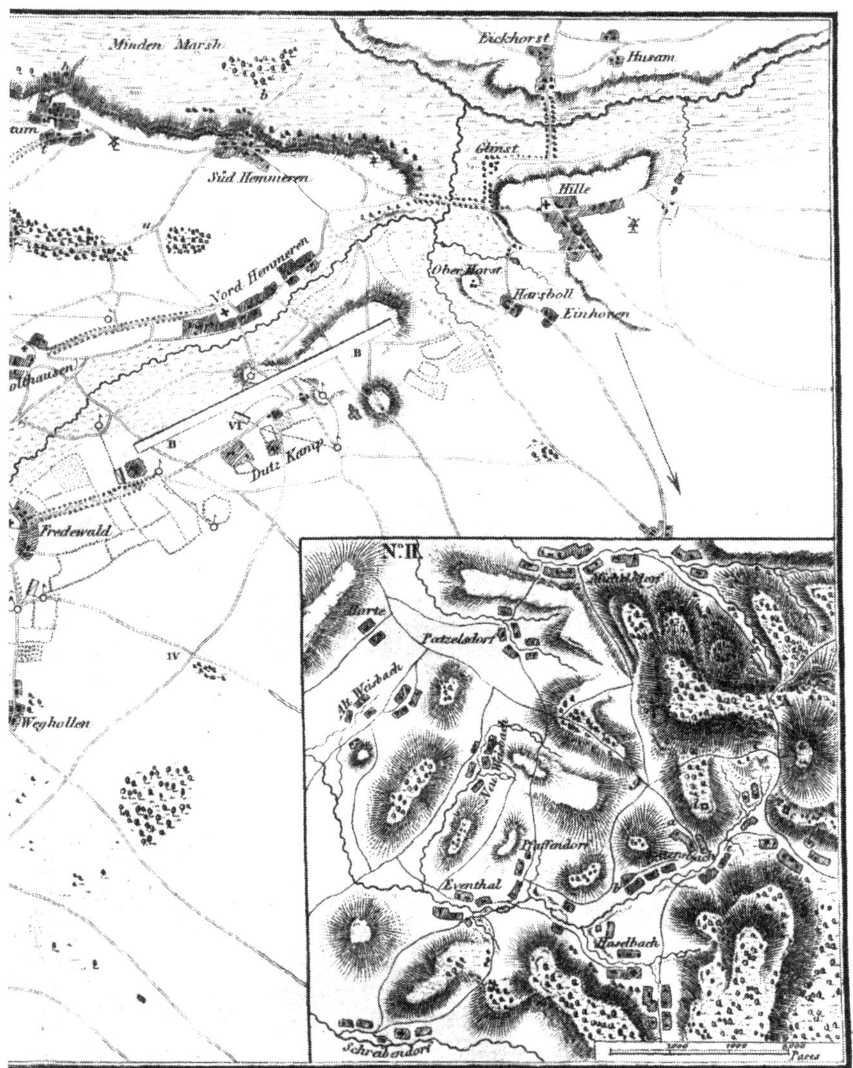

ilitary Library Whitehall.

Published August 30th 1831 by T. Ege

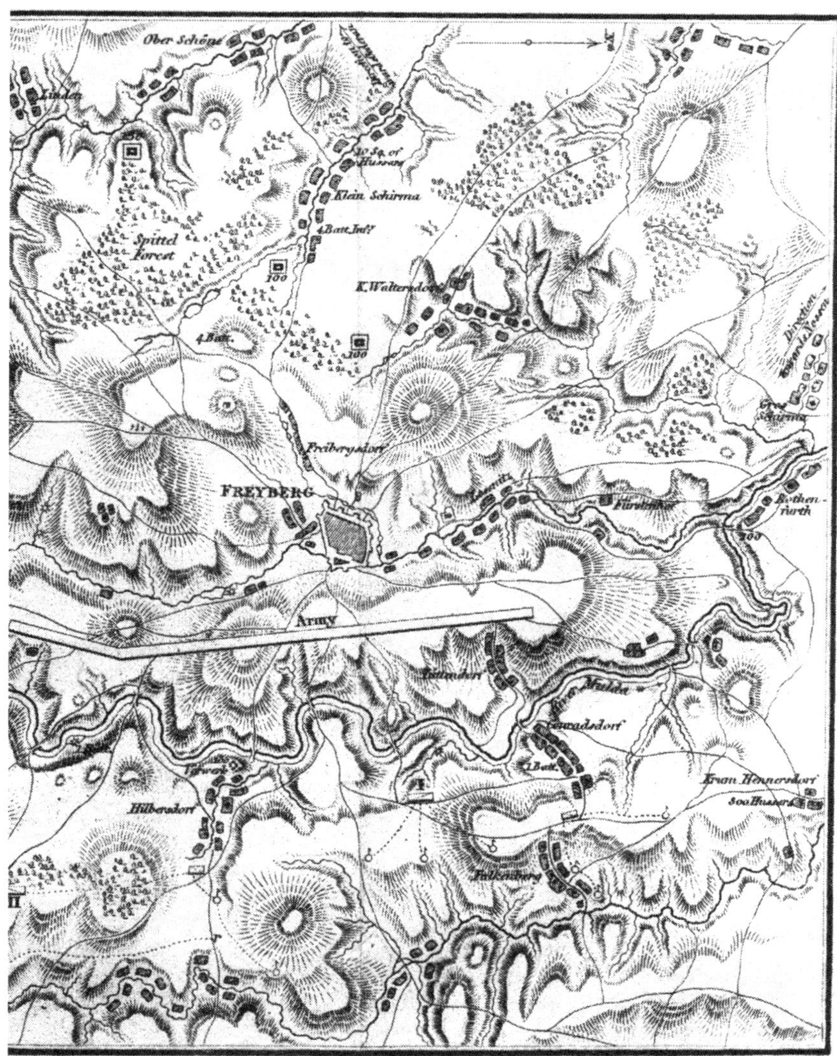

Drawn & Engraved by I. Hofmann & C. Marquard.

...rton, Military Library Whitehall.

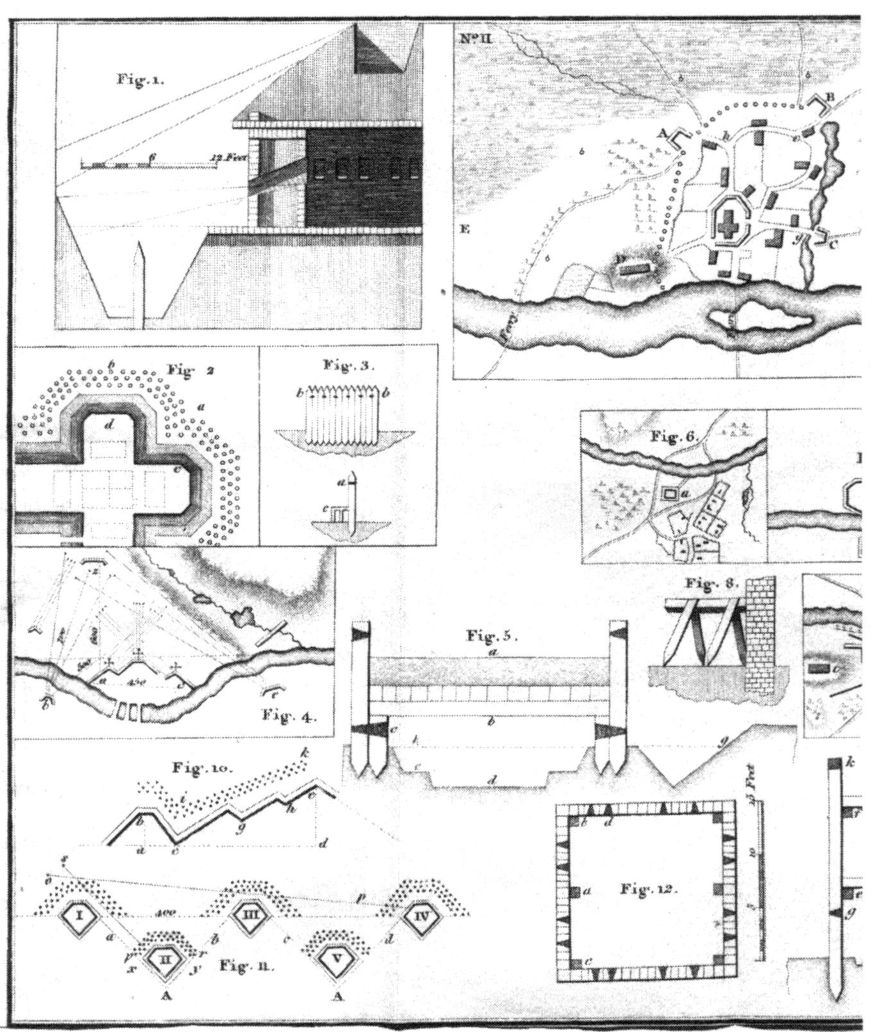
Fig. 1.
12 Feet
No. II
Fig. 2
Fig. 3.
Fig. 4.
Fig. 5.
Fig. 6.
Fig. 8.
Fig. 10.
Fig. 11.
Fig. 12.
15 Feet

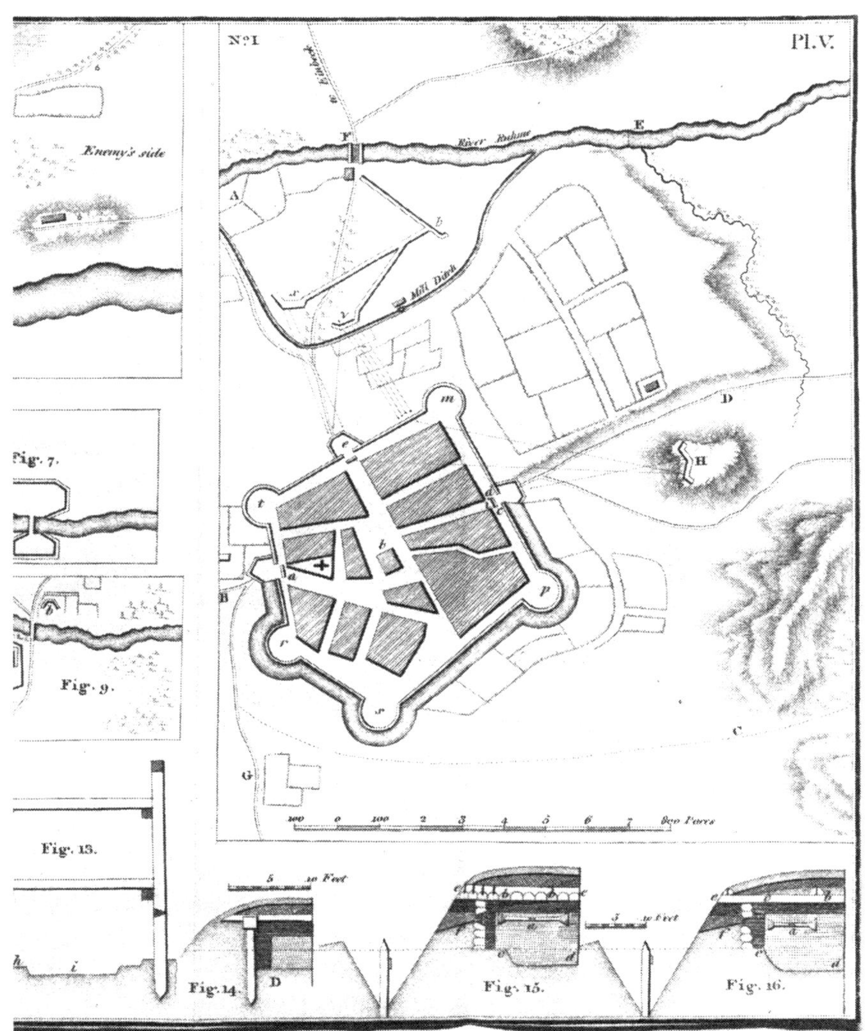

Nº I
Pl. V.
Enemy's side
River Ruhme
Mill Ditch
Fig. 7.
Fig. 9.
Fig. 13.
Fig. 14.
Fig. 15.
Fig. 16.
A
B
C
D
E
F
G
H
100 0 100 2 3 4 5 6 7 800 Paces
5 10 Feet

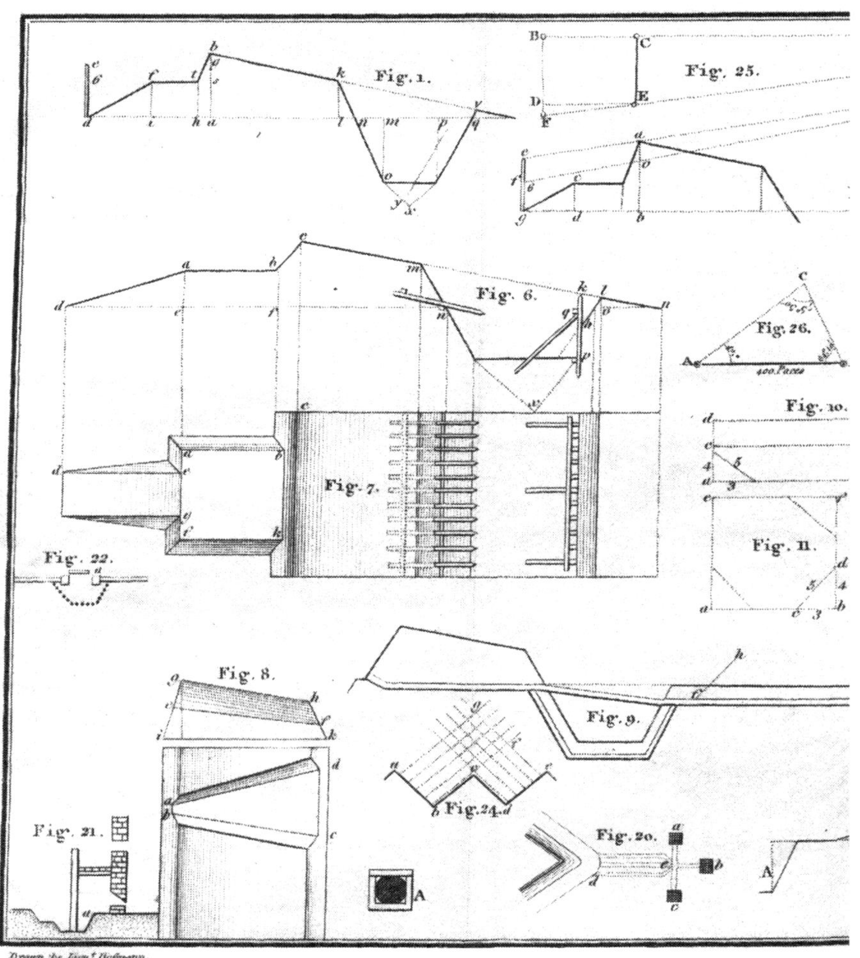

Drawn by Lieut. Hofmann.

Published August 30th 1811, by T. Egerto

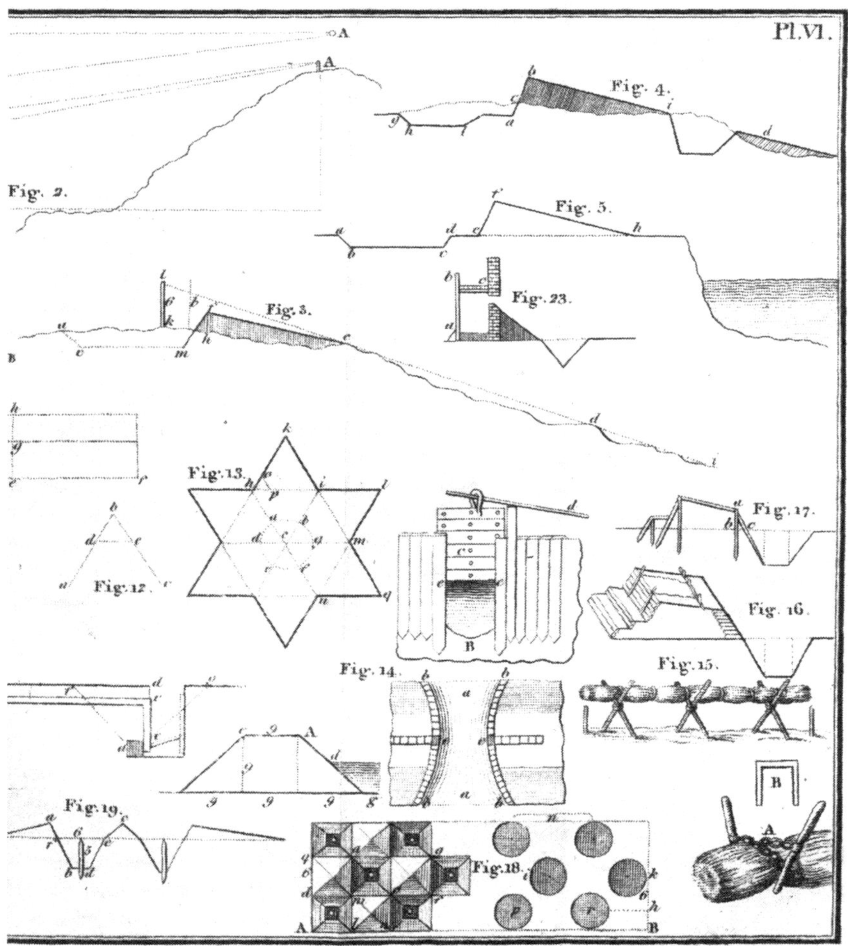

Engraved by C. Marcuard.

n Military Library Whitehall.

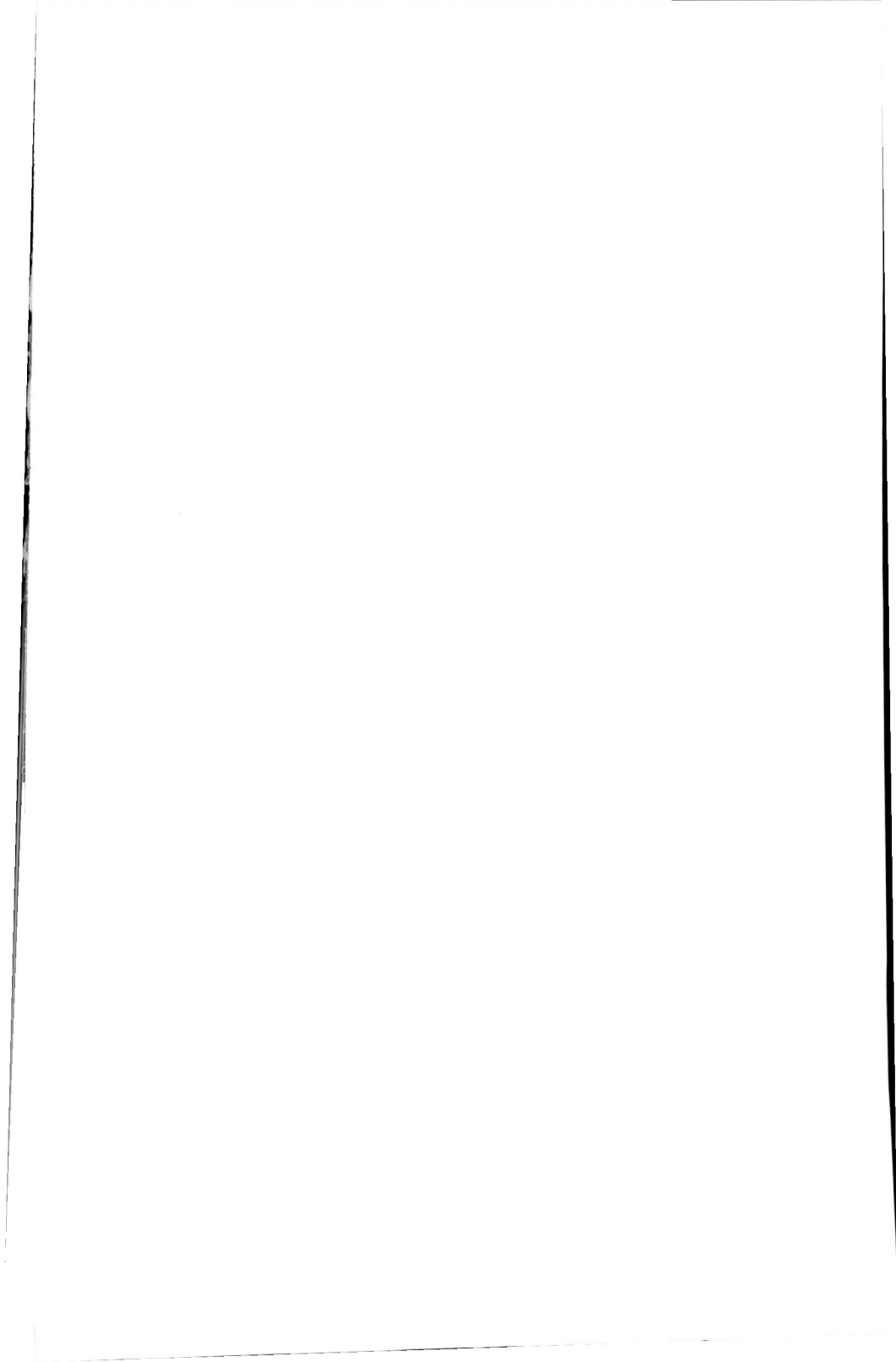

www.ingramcontent.com/pod-product-compliance
Lightning Source LLC
Chambersburg PA
CBHW031402180326
41458CB00043B/6579/J